S0-BKS-269

THE SCIENCE OF WINE

Cedric Austin

AMERICAN ELSEVIER PUBLISHING COMPANY, INC.
NEW YORK 1968

AMERICAN ELSEVIER PUBLISHING COMPANY, INC.,
52, Vanderbilt Avenue,
New York, N.Y. 10017

Library of Congress Catalog Card Number 68-26815

Printed and bound in Great Britain

Preface

The French have a saying that good wine results from the happy marriage of art and science. Winemaking is by nature a craft calling for skill and experience, but the changes that take place in vat and cask are chemical reactions, and the interpretation of these is a science. The modern vintner, therefore, is a blend of craftsman and scientist, one who is continuously adapting the traditions of the past in the light of progressive scientific knowledge.

The difficulty in understanding the scientific nature of wine and its structural properties is that this involves many highly specialised fields; one needs to go beyond general works on organic chemistry and biochemistry to more specific confines such as mycology, oecology, cytology, the classification of yeasts and formation of glycosides. Many textbooks are available on these somewhat esoteric subjects intended for the undergraduate or research worker; they are seldom aimed at the simpler and more practical needs of the vintner. Without some external help the enterprising student of winemaking becomes bogged down in irrelevant material, even though he may have access to a specialised library.

This book is an effort to surmount such obstacles by showing concisely the part played by science that applies to wine and winemaking, and presenting it in a way comprehensible to the more conscious reader. The volume has four sections describing the main constituents of wine: yeast, sugar, alcohol and acid, and a fifth covering wine disorders. More advanced topics, such as stereochemistry of sugars and the glycolytic sequence, are described in the sixth section. Formulae have been presented throughout the book.

Finally, I am very grateful to my friend and fellow-winemaker Dr Bernard Archer, B.Sc., A.R.I.C., who read through my typescript for scientific accuracy of detail and made constructive suggestions to improve its final form.

Contents

To all those with whom I have taken wine

Section 1 Yeasts

Chapter 1 Yeast Cytology and Reproduction

The Cell

The cells which make up a lump of yeast are minute (5–7μ in diameter: $1\mu = \frac{1}{1000}$ mm, or roughly $\frac{1}{25000}$ inch) and are round, oval, elongated, pointed, according to the species. In view of their extremely small size, there exists much confusion about the identification of cellular components, such that different authorities variously apply the terms nucleus, centrosome and nucleolus, for example, to the same structure.

The following diagram of a yeast cell is based on that published by Lindegren (1952):

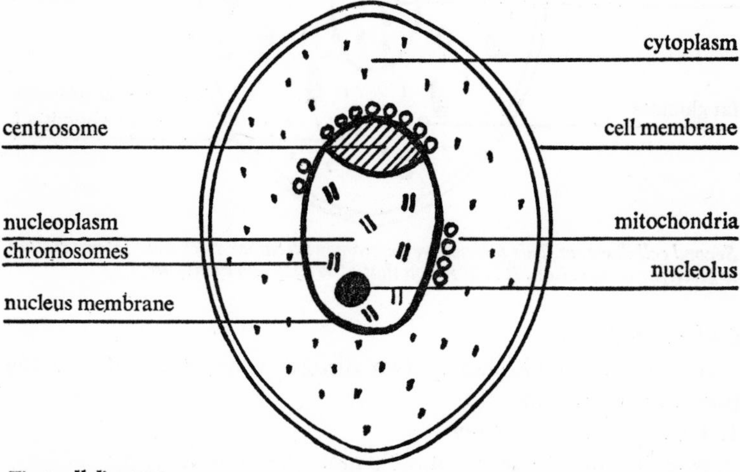

First cell diagram

Each cell is surrounded by a transparent semi-permeable membrane which permits small molecules to diffuse in and out. Within the cell wall is the protoplasm, a viscous colloidal protein solution in which are one or more sharply differentiated vacuoles containing an aqueous solution of, in the main, non-proteinaceous material. Embedded in the protoplasm is the nucleus, enclosed by a nuclear membrane and composed largely of nucleoplasm. Special structures within the nucleus are the nucleolus, a dark or shining body which probably represents a nucleic acid reservoir, and the centrosome, which is possibly implicated in cell reproduction. The nucleus has a controlling influence on the synthesis of cellular protein and is responsible to a large degree for the transmission of hereditary characters from generation to generation. The protoplasm also contains other much smaller particles, of which the most obvious are the mitochondria, representing the site of certain enzymic activities and responsible for the final reactions of aerobic respiration.

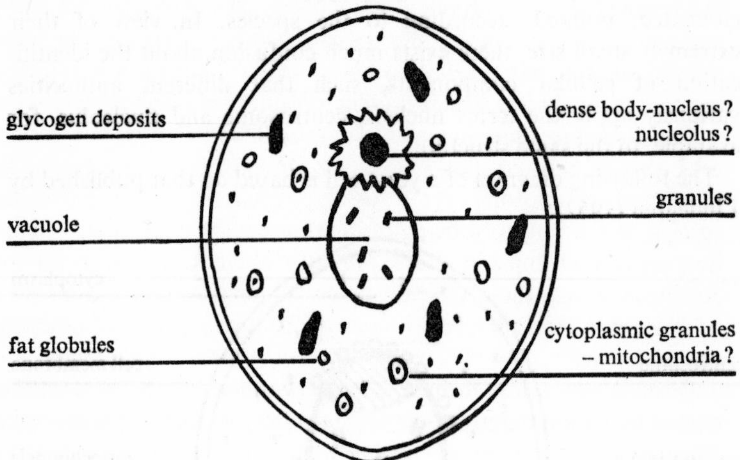

Second cell diagram. This alternative diagram, with rather a different interpretation, is given for comparison. It is based on that of Wager and Penistone.

Cell Reproduction

A yeast cell multiplies in two distinct ways, according to the prevailing conditions.

1. *Vegetative Reproduction*

When conditions are favourable most yeasts multiply by budding. Favourable conditions means:

a. Abundant supply of appropriate nutrients
b. Suitable temperature
c. Moist environment.

Buds originate as local swellings on the cell surface, and their position may be distinguished as bipolar when they arise at the ends of elongated cells, or multipolar when there is no obvious orientation. Continued growth of the bud and progressive constriction at the neck where it joins the mother cell leads ultimately to abstraction of an independent daughter cell. In some species failure of the daughter cell to break away leads to the formation of a pseudomycelium composed of clusters or chains of cells.

MITOSIS OR DIVISION OF SOMATIC CELLS. While this is happening externally, very important events are taking place inside the cell. The nucleoplasm of all cell nuclei contains a special material called *chromatin*. The name was given to it by Flemming (1843–1905) because of the ease with which it could be stained. When the cell is in a resting (i.e. not reproductive) condition the chromatin frequently appears to form a complex and delicate network extending throughout the nucleus, but as cell division becomes imminent it is gathered together in thread-like structures which Waldeyer named *chromosomes*.

Chromosomes differ in length and general appearance, but normal cells contain two recognisably similar individuals of each type. The number of chromosomes is a constant characteristic of the particular species, but owing to difficulties of observation there is still doubt how many are formed in yeasts: estimates range from four to twenty. Arranged along these chromosomes, somewhat like beads on a necklace, are the *genes*, which control the development of inherited characteristics.

When a new cell is formed, a process called *mitosis* (a term also coined by Flemming, from the Greek word for 'thread') takes place. The homologous pairs of chromosomes split along their lengths, each chromosome thus forming two *chromatids*, held together at a point called the *centromere*. The chromatids then separate and move in two similarly constituted groups to opposite poles of the nucleus. Formation of a nuclear membrane between the two groups give two identical nuclei, one of which migrates into the bud and becomes the nucleus of the daughter cell. This is a much condensed account of what is usually represented as a succession of different 'phases', but it indicates the manner in which the daughter cell, in virtue of its identical chromosomal content, inherits in every respect the characteristics of the mother cell.

It is interesting to note that the fruit fly, *Drosophila melanogaster*, the bane of the winemaker's existence, has involuntarily done a great deal to aid research in this direction. Their rapid reproduction rate, combined with the fact that their cells contain only four pairs of readily distinguishable chromosomes, made them an important tool in genetical research, and work with Drosophila formed the basis of Morgan's development of the gene theory.

Such are the events taking place inside the yeast cells as they reproduce by vegetative growth during fermentation. The rate of growth varies considerably according to the efficiency with which oxygen is excluded; this is discussed in chapter 2.

2. *Sporulative Reproduction*

Of the three yeast families, one, the so-called 'pseudo-yeasts', forms no spores at all, and another ejects its spores as projectiles. We are, however, concerned only with the third family, Endomycetaceae, in which in common with other Ascomycetes, spores are formed in a cell called the *ascus*. The following description refers to the typical genus *Saccharomyces*.

As autumn comes to an end, the fruits disappear and the supply of nourishment dries up; at the same time the contents of surviving yeast cells become differentiated into a series of spores, each surrounded by its own cell wall and all enclosed by the persistent walls of the original cells, which have now become *asci*. The term *ascus* means 'sac', and these particular spores are ascospores. Such spores, which are functionally equivalent to the complex seeds of higher plants, represent a resting stage in the yeast life-cycle. Due, it appears, to their low moisture content and impermeable walls, they can survive conditions under which normal cells would perish. When conditions improve they take up water, swell and become ordinary yeast cells; some of them, carried perhaps by the wind, insects or birds, find their way on to fruits, where they once again multiply by budding. Yeast cells can be induced to form spores in the laboratory by growing them in a nutrient medium, and then transferring them to an environment deficient in nutrient, such as gypsum blocks, washed agar or moist filter paper.

It was as far back as 1891 that Hansen noticed that ascospores were able to fuse together at germination, but any idea that yeast cells might behave sexually was then unacceptable; even in 1928, Guilliermond was still maintaining that yeasts had lost all traces of sexuality. It is due largely to Winge, a later follower of Hansen at the Carlsberg Laboratory, who between the wars did much research on the life cycle of yeasts, and also to Lindegren for his University research on the same subject, that we are now able to attach more significance to ascospore formation.

MEIOSIS OR REDUCTION–DIVISION OF GAMETAL CELLS. It was shown above that the number of chromosomes in an ordinary somatic (Greek *soma* = body) cell is a constant characteristic of the species, but there is an important exception to this: the sex cells, known as *gametes*. It is obvious that these could not contain the full number of pairs of chromosomes, for if this was the case the number would be doubled each time that a male and female gamete fused in the act of fertilisation, and this is not so. When sex cells are formed a process called *meiosis*, from a Greek word meaning 'reduction', takes place. In this the paired chromosomes separate, one member of each pair moving to opposite poles of the nucleus, so that when the nucleus subsequently divides into two each new nucleus contains half the usual chromosomal complement. This is the reduction–division that characterises meiosis.

Following this initial division, each nucleus divides again in an essentially mitotic manner, each chromosome separating into chromatids, to give four nuclei, each with half the usual chromosome number. Formation of cell walls round each nucleus and its associated cytoplasm results in the appearance of four ascospores enclosed by the wall of the original cell.

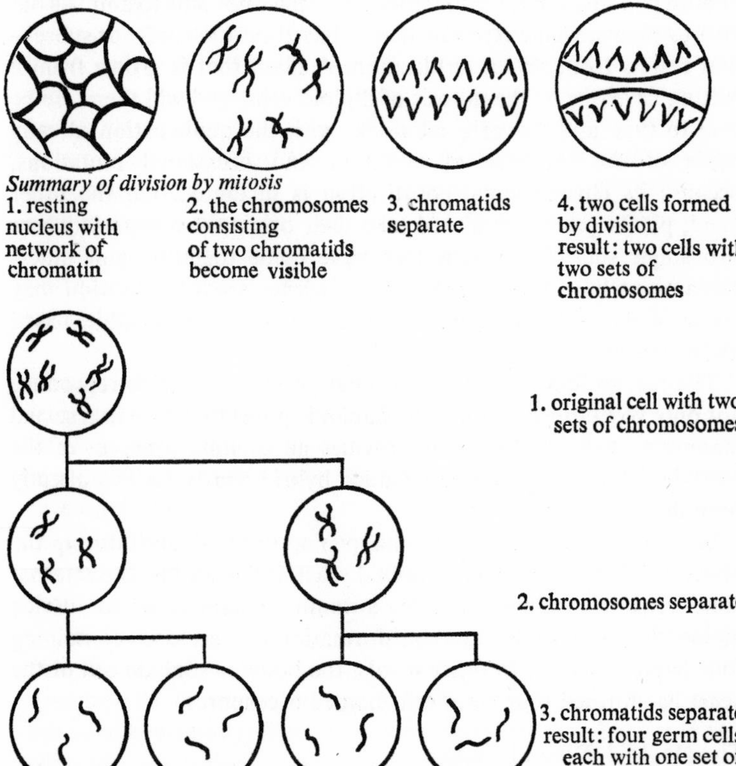

Summary of division by mitosis

Summary of division by meiosis

Splitting of chromosome during mitosis
A pair of chromosomes is thus formed each of which carries a portion of every constituent section of the chromosome

Cells which contain two sets of chromosomes, i.e. the number found in somatic cells, are called *diploid* cells; those with half this number, i.e. only one from each pair, are known as *haploid* cells. Thus diploid human cells contain 46 chromosomes, or 23 pairs, or two sets; haploid human cells contain 23 single chromosomes, or one set. When fertilisation takes place and the male haploid spermatozoon fuses with the female haploid egg, then a diploid cell (*zygote*) is formed in which the normal chromosome number is re-established.

It is obvious that this intermingling of two different single sets of chromosomes may produce an offspring with mixed characteristics of both parents, and the same situation applies not only to animals but also to plants, where reproduction is based on the seed—or spore—that results from sexual fertilisation. A plant that is grown from a cutting develops by the mitosis of diploid cells, and will therefore be true to type and the original stock, with the qualification already made of the principle of variation and occasional mutations, because its chromosomal constitution is unaltered. On the other hand, plants grown from seed take their origin from zygotes where two single sets of chromosomes have come together, and unless precautions have been taken by the gardener cross-fertilisation may occur. Where controlled, interesting hybridisation is possible, based on the process of meiosis.

The recognition in recent years that yeasts are able to reproduce not only vegetatively but also by haploid spores that fuse in a sexual manner to form diploid cells provides an exciting prospect of the cross-breeding of species to produce hybrid yeasts, as has already been done with vines.

We can now sum up what is happening inside the cell during the process of forming spores. A diploid yeast, influenced by the external conditions prevailing in late autumn, commences to divide meiotically. This leads to its transformation into an ascus containing four haploid spores. In other words, the body or somatic cell of the yeast has formed four germ cells named ascospores.

The Haploid–Diploid Cycle

The sexual conjugation of yeasts may take place between spores while they are still contained in the ascus, or later, after germination,

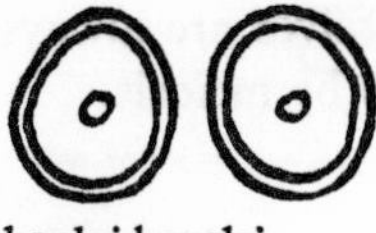
haploid nuclei

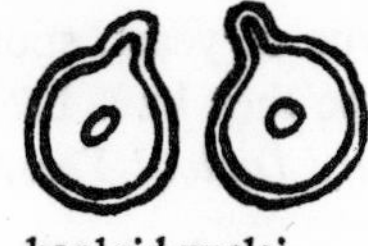
haploid nuclei

diploid nucleus

Spores fusing

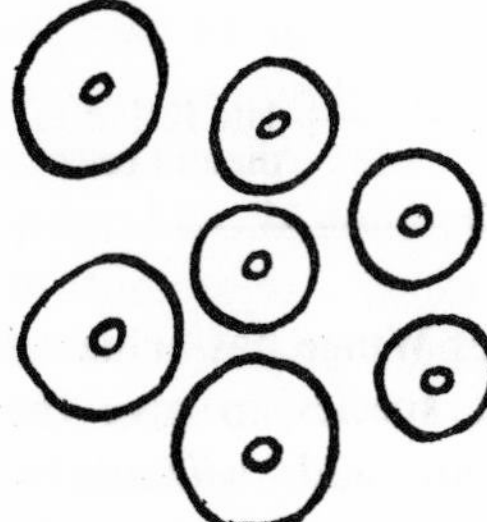
haploid nuclei

diploid nuclei

Cells fusing

between haploid vegetative cells. Lindegren considers that two of the four spores are of one 'sex-type' and the other two of an opposite 'sex-type', similar mating types being unable to combine. If fusion occurs between two pairs of haploid spores within the containing wall, then two diploid cells will be formed, each capable of producing diploid colonies by budding in the usual way. When conjugation does not occur at this stage the spores can germinate and form haploid colonies: haploid vegetative cells are usually smaller.

Yeasts vary markedly in the extent of their haploid existence and can be classified according to whether their vegetative cells are predominantly haploid or diploid. The haploid cells of *Saccharomyces cerevisiae* bud only a few times before conjugation takes place, so this yeast is described as a diploid species. The distinction is more obvious in some cases than in others: where conjugation occurs within the ascus, as in *Saccharomycodes ludwigii*, the vegetative cells are invariably diploid. In some yeasts (*Schizosaccharomyces octosporus*, *Saccharomyces* spp. formerly referred to as *Zygosaccharomyces*) conjugation occurs between vegetative cells and is immediately followed by ascus formation, so that no diploid colonies are ever produced. Some yeast genera have lost entirely their ability to produce spores, and thus the genus *Torulopsis*, for example, can be regarded as a form of *Saccharomyces* which has become permanently haploid. Whatever the type of yeast, spores themselves are always haploid, of course, because they have been formed by meiosis.

Life Cycle of Predominantly Diploid Yeast

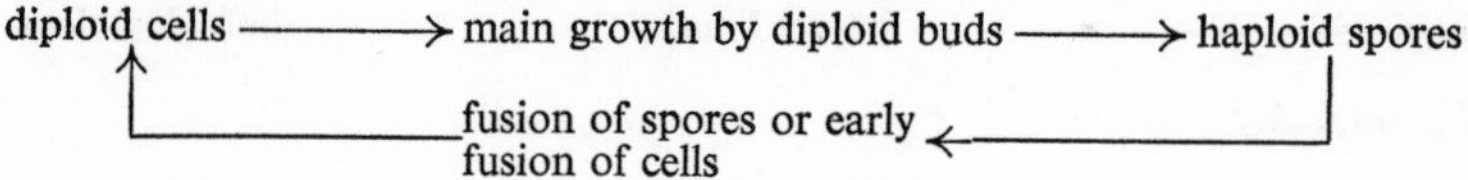

Life Cycle of Predominantly Haploid Yeast

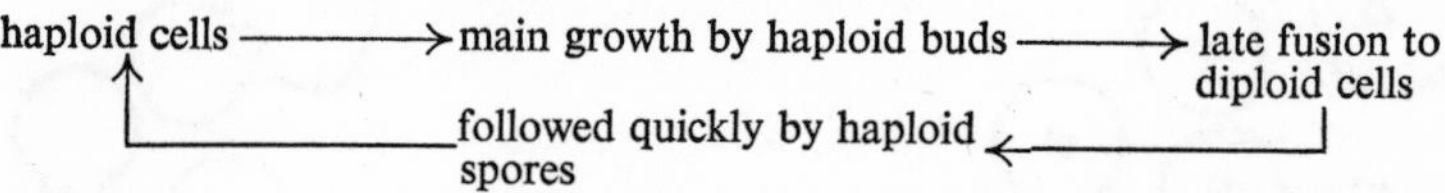

By the use of micro-manipulative techniques Winge has achieved conjugation between ascospores of different species to give the hybrids *Saccharomyces elipsoideus* × *willianus* and *willianus* × *italicus*. This production of hybrids opens exciting avenues to the development of new strains.

Variation of Inherited Characteristics

The process of mitosis ensures transmission of those hereditary characters which constitute the genotype of the species. Individual representatives of any species are, however, not all the same, and it is obvious that some degree of variation can occur.

1. GENE SEGREGATION AND RECOMBINATION. The two haploid cells involved in conjugation are unlikely to have the same genetic constitution; although this might be the case when conjugation takes place between related spores inside an ascus, it is beyond the bounds of probability when vegetative cells from different colonies are concerned. This means that the zygote acquires a set of genes differing from that of the foregoing diploid generations, and if the differences are sufficiently important the new generation may show perceptible variation from the 'type', at least in some respects.

This aspect can be illustrated by reference to a person with brown eyes, who may have had a brown-eyed mother but a blue-eyed father. In this case he will carry genes for both blue and brown eyes, one on each of the appropriate pair of homologous chromosomes. Since these chromosomes are separated at meiosis, half of his reproductive cells will have the blue factor and the other half the brown. The gene for brown eye colour is dominant, and blue eyes appear only in children who have received the blue gene from both parents.

When it is considered that all the gene-controlled characteristics of a cell are subject to this sort of segregation and recombination in a semi-independent manner it is easy to visualise that there will be marked variation from individual to individual, even among yeasts.

2. GENE MUTATION. Very occasionally, a gene is modified in such a way as to produce a completely new effect. This change (mutation) occurs spontaneously and is not predictable; it is only detectable if it results in an obviously different appearance or metabolic activity, and a vast number of mutations probably go unnoticed.

In general, mutations which result in major deviation from the average type will be either beneficial or harmful. In so far as they represent a better adaptation to the environment, they are likely to be more successful and thus affect the ultimate characteristics of a culture by the process of natural selection.

Induced variations can be initiated by treatment with ultraviolet light or X-rays, which increases the rate of mutation about 150 times, but the changes produced are still completely unpredictable.

3. ENVIRONMENTAL MODIFICATION. The character and behaviour of an organism is influenced by its environment, but since its genetic constitution remains unchanged, the acquired characters usually persist only as long as the environmental conditions remain the same.

Yeasts show many instances of environmental modification, and it is well known that they may be 'trained' to accept high concentrations of sugar, withstand high temperatures, increase in alcohol tolerance and so on. When trained over a long period the modifications may become more stable, persisting after the environment has become changed. The term *dauer-modification* has been suggested for this from the German word *dauern* = 'to endure'. However, the effect of environmental change on the distinctive properties of a particular yeast strain has raised doubts on the use of topographical strains of *S. cerevisiae* var. *ellipsoideus* to confer the properties of a certain type of continental wine on a home-made product. If a 'Beaune yeast', for example, varies sufficiently in its characteristics from other ellipsoidal wine yeasts to be regarded as a distinctive *strain* (it is certainly not acceptable as a different *variety*), it is because it has been trained over centuries to the grape juice that is peculiar to the vineyards of Beaune in Burgundy. Any modification in its character is environmental and not apparently hereditable. If used to ferment, say, elderberry juice the constant environment of the must is changed with respect to sugars, acids, tannins, salts and other constituents; it is doubtful how long the acquired character of the Beaune yeast will persist, and how quickly this will become modified by the new environment so that the yeast changes into what could be termed an 'elderberry strain'. There is at present a regrettable lack of information and research on this subject.

4. DETERIORATION OF CULTURES. In recent years two authorities have emphasised the tendency of yeast sub-cultures slowly to deteriorate, on the one hand (Fowell), because of an accumulation of undesirable mutants during prolonged vegetative multiplication, and on the other, because of the appearance of extreme variants as a consequence of sporulative reproduction. Associated with the latter aspect, Winge argues that only a haploid culture derived from a single spore can retain its special characteristics through successive sub-cultures.

Chapter 2 Yeast Fermentation

The History of Fermentation

At the beginning of the last century the general opinion of chemists was that fermentation was a purely chemical process. The French chemist Antoine Lavoisier, who was guillotined in the French Revolution on the trumped-up charge of having watered the soldiers' tobacco, had decided that alcoholic fermentation resulted from a chemical splitting of the sugar molecule. He based this on his observations that the resulting qualities of alcohol and carbon dioxide gas formed were each roughly half the amount of sugar fermented. He was remarkably correct in this, but his deduction that this was a chemical reaction unconnected with any form of life was quite mistaken.

About half a century later, in 1837, another Frenchman, the physicist Charles Cagniard de la Tour, reported his observations on fermentation. He had closely inspected drops of fermenting beer-wort under the microscope, and noticed yeast cells forming buds. The similarity of the process with sprouting seeds convinced him that the cells were alive, and he reached the conclusion, in advance of his time, that fermentation was the result of yeast growth and therefore part of a living process. The chemists of the day found this thesis quite unacceptable, but their main difficulty in refuting it lay in explaining away the presence of yeast. It was one thing to say that yeast appeared as a result of spontaneous generation, just as maggots in cheese and worms in dung were supposed to be formed, but why was its presence always correlated with this fermentation process?

The great Swedish chemist, Jöns Berzelius, 1779–1848, who was also responsible for the idea of representing elements by their initial letter, had a new idea. He had recently coined the term 'catalyst', from the Greek word *katalysis*=dissolution, for any substance which by its presence accelerated the reaction of other substances without itself undergoing any change in the process. (Thus a mixture of hydrogen and oxygen can be kept indefinitely at room temperature, but introduce platinum black as a catalyst and they combine at once to form water, without any effect on the catalyst.) Berzelius now included yeast as a catalytic agent which by its mere presence in a solution of sugar caused a breakdown into alcohol and carbon dioxide, and not by taking any biological part in the reaction. The yeast itself was a 'ferment' (used in the sense of 'enzyme', but this word had not been coined) produced by the spontaneous oxidation of grape juice.

He was backed up by another great contemporary chemist, Justus von Liebig, 1803–73, the first to divide food into fats, carbohydrates and proteins. He, too, deprecated the biological point of view and the value of microscopic evidence, but he put forward an alternative to the catalystic theory of Berzelius. He linked fermentation with putrefaction, and declared that it was the decomposition of yeast that produced the fermentation of sugar into alcohol and carbon dioxide. The yeasts, he argued, died and decomposed, thus forming albuminous material, the rotting of which caused the fermentation. It seems curious today that such a distinguished and progressive chemist was so blind to the facts of fermentation, but one must remember that the organic chemistry of carbon-containing compounds really began in 1828, when Hennell prepared alcohol synthetically from ethylene and Wöhler produced urea artificially in the laboratory; chemists were impatient of anything explained on biological or *vitalistic* grounds as actions that put the clock back.

It was at this stage that Louis Pasteur came on the scene, a French chemist by training, but eventually the world's first bacteriologist. He was born in a poor family in 1822, and trained as a teacher. In 1847 he took his doctor's degree in science, and in 1854 he was appointed Professor of the Faculty of Sciences at Lille. The father of one of his students there, a certain M. Bigo, came to him with a problem. He manufactured alcohol from beetroot, and his fermentations were going sour, involving him in considerable financial loss. When Pasteur examined samples under the microscope he could distinguish long, thin, rod-like organisms quite different from the oval and round cells of yeast. Plainly these must be the cause of the soured must, since they were absent from sound samples. Were they another form of yeast infecting the wine? Pasteur had already done research work on tartaric acid and on amyl alcohol, and suspected that fermentation was connected with living organisms. Now in 1855 he began his fermentation studies in earnest.

His first new studies were with the souring of milk resulting from lactic acid production, and perhaps his choice of this medium in preference to wine was because Liebig in his attack on the supporters of the biological point of view had pointed out that in 'milk fermentation' certainly no yeast was necessary. ('Fermentation' is used here in the sense of 'souring', not in the sense of converting lactose to alcohol.) By examining the deposit of such souring, he found tiny figure-of-eight cells, and not only did these live and multiply but they were also able to produce lactic acid 'fermentation' in sound milk. Pasteur regarded them as lactic acid yeasts, whereas we know they are bacteria, but nevertheless he had shown a *vitalistic* cause for the change, as opposed to Liebig's view.

Next he turned to alcohol-fermentation, and Liebig's views on putrefaction. He discovered that yeast would grow on a medium containing sugar, ammonium tartrate (a source of nitrogen), phosphates and the inorganic residue from ashed yeast. This mixture clearly contained nothing which could putrefy and so cause fermentation, yet a yeast inoculum the size of a pin-head grew, budded and fermented the sugar. Liebig refused to accept this evidence, but the vitalistic basis of fermentation had been established.

In 1860 Pasteur published his *Note on Alcoholic Fermentation*, a landmark in our knowledge of this process. In it he stated, 'The chemical act of fermentation is essentially a process correlative with a vital act, commencing and ceasing with the latter. There is never alcoholic fermentation, properly so-called, without organisation, development and multiplication of cells or without the continued life of cells already formed. The results expressed in this note seem to me to be completely opposed to the opinions of Liebig and Berzelius.' It is difficult today to appreciate the tremendous task Pasteur was facing in endeavouring to prove his thesis. One of his biographers, René J. Dubois, says of this, 'In 1857 the chemist who adopted the vitalistic theory of fermentation had to face the same odds that would today confront a telephone engineer interested in developing the use of telepathy for the transmission of thought'. Further research work followed, and Pasteur produced his 'Studies on Beer', in which he showed that spoilage of wine and beer was due to the action of micro-organisms present in the air or on contaminated apparatus. He also demonstrated that no changes occurred in a medium if it was first heated to 125°–140° F, a process now called pasteurisation. It is no exaggeration to claim that bacteriology started with the study of fermentation.

Liebig held out till the end of his days, but the balance was in favour of the vitalistic theory. Yet Pasteur was not entirely correct, and he seems to have felt this himself. When Traube suggested that an organism might produce a chemical substance that acted as a fermentation catalyst, Pasteur was willing to agree. But he pointed out that it made little difference to his ideas unless fermentation could be produced in a cell-free system. Consequently he distinguished between 'organised ferments' that are closely connected with living cells and 'unorganised ferments', such as diastase and pepsin, which are active apart from living cells.

In 1897 two German chemists, the brothers Eduard and Hans Büchner, were working on yeast with the intention of providing an extract that would be of medicinal value. It proved possible to make an extract by grinding yeast with sand and kieselguhr (a silicaceous earth) and squeezing the product under a hydraulic press. The difficulty was now the question of preservation. One possible method was the addition of sugar (sucrose), this idea probably being suggested by its use in the kitchen for jam. To their astonishment, they found that the mixture of juice and sugar fermented, and that alcohol and carbon dioxide were formed. Quite by accident, and with another purpose entirely in mind, they had stumbled on a remarkable discovery. A juice obtained from yeast, *but containing no intact yeast cells whatever*, had fermented sugar. Wilhelm Kühne in 1878 had coined the word 'enzyme' for that 'something in yeast' that caused fermentation, from Greek *en*='in', *zume*='yeast', and this was later used as a collective name for organic catalysts, the particular enzyme in yeast responsible for fermentation being named *zymase*. We now know that 'zymase' is not a single enzyme, but a complex mixture of several enzymes, together with their related *coenzymes*, and it is usual to refer to the 'zymase complex'. Finally, after further experiments, the Büchners were able to claim, 'The production of alcoholic fermentation does not require so complicated an apparatus as the yeast cell.' In 1907 Eduard received the Nobel Prize 'for his discovery of cell-less fermentation'.

At first sight this looks like the end of Pasteur's vitalistic theory of fermentation, but in fact it brought together the two opposing theories that had been battling during the century. The change of sugar to alcohol is indeed a chemical process due to lifeless substances called enzymes. But these enzymes can be produced only by living organisms, despite efforts to manufacture them in the laboratory. So in the light of Büchner's discovery, we can bring the chemical and biological views together by modifying Pasteur's 'No fermentation without life', to 'No fermentation without enzymes; no enzymes without life'. We can now turn our attention to the enzymology of fermentation.

Enzymes

It was back in 1833 that two French chemists, Payen and Persoz, isolated a substance from malt that was capable of converting starch to sugar, and named it *diastase*, from the Greek *diastis* = 'separation'. A little later, Schwann isolated the protease *pepsin* from gastric juice. Both of these preparations were enzymes, although of course the collective name for them was not coined for another fifty years.

The discovery of the Büchner brothers was soon the subject of international research, and Harden and Young added to our knowledge of 'zymase', revealing some of the characteristics of enzymes in general. They found, for example, that zymase consisted of two component parts, an enzyme fraction and a coenzyme fraction. Many enzymes function only in the presence of non-protein components, relatively small molecules, which may be tightly bound to the enzyme itself as 'prosthetic' groups, or more loosely connected as coenzymes. Many others require only metallic ions, which are usually rated as 'cofactors' but not coenzymes. The complete enzyme–coenzyme complex contributes the *holoenzyme*, the protein constituent being the *apoenzyme*. This latter activates the substance upon which it works and is involved in its absorption; the coenzyme then transfers atoms from this activated molecule to another molecule, thus completing the reaction, the cycle repeating itself again and again. Cozymase (coenzyme I) was the first known coenzyme, although it was not isolated until 1936 by von Euler; Sir Alexander Todd synthesised it in 1956.

Harden and Young discovered that cozymase could be separated from the zymase complex by *dialysis*. When a semi-permeable sac containing yeast juice was suspended in water all the small molecules, including the coenzymes, diffused out, leaving the large protein molecules behind. Since zymase is inactive in the absence of cozymase, this resulted in the loss of fermentative activity; but activity could be restored if the diffusate was concentrated and returned to the sac contents. Boiled, otherwise inactive yeast juice was also equally effective in restoring the activity of the dialysed protein fraction. Harden received the Nobel Prize jointly with von Euler in 1929 for his work on enzymes.

Enzymes may be defined as 'organic or biological catalysts', and we have already mentioned that a catalyst is a substance that accelerates a chemical reaction without being itself changed or consumed in the process. For example, in the manufacture of esters by the reaction of acids upon alcohols, the time taken is reduced from months to hours by the addition of a suitable catalyst. In *theory*, catalysts can only accelerate the progress of reactions, and these reactions must be capable of taking place spontaneously without them. In *fact*, such reactions may in certain cases be so inordinately slow that from a practical point of view they may be regarded as not really taking place at all in the absence of a catalyst. To clarify the part they play in reactions, catalysts have been compared with lubricants, but a better idea is gained by regarding them as foremen sent to take charge of a gang of workmen; little is being done until the foreman arrives on the scene and then, without taking part in the work himself, he soon sorts things out and gets the work speeded up.

Enzymes differ in two important respects from inorganic catalysts. In the first place they are invariably products of cellular metabolism; secondly, they generally show a high degree of specificity and catalyse only one reaction, or reactions of a particular type. Generally they are named by adding the suffix *-ase* to the name of their substrate, i.e. the substance affected by them. Thus *maltase* acts on maltose, *protease* on protein and so on. Where there are exceptions to this rule, the names were given at an early stage, as with *ptyalin*, the old term for salivary amylase.

A remarkable feature of enzymes is the colossal amount of work done by a very small quantity of enzyme. *Invertase*, the sucrose-splitting enzyme, will convert one million times its weight of sugar without loss of activity. They are, however, *thermolabile*, and are irreversibly inactivated if they are held at 60° C for more than a short time, or if they are heated to 100° C. Otherwise, under less extreme conditions, enzymic activity increases with increase of temperature; enzymes are inactive below O° C. Dry enzyme preparations are frequently more resistant to heat inactivation than are enzyme solutions.

Each enzyme has an optimum temperature and pH at which they give their best performance. Beermakers will know how slight a variation in mashing temperature affects the α-amylase and the β-amylase, encouraging one of these at the expense of the other, and so altering the final character of the brew. Not only are enzymic reactions extremely sensitive to temperature and acidity or alkalinity but also to metallic ions and often to the reaction products.

Some 650 enzymes have now been isolated, more than 120 of them in crystalline form, and the number is continually increasing. The first crystalline enzyme, urease, was obtained by Sumner in 1926, just a hundred years after the synthesis of its substrate, urea, by Wöhler had initiated organic chemistry as such and refuted the vitalistic theory that organic compounds could be made only by living organisms.

The enzymes which are of special importance in winemaking can be arranged in four groups:

Sucrolytic: invertase breaks down sucrose, and *maltase* degrades maltose into fermentable forms.
Glycolytic: the *zymase* complex produces the long chain of reactions whereby sugar is converted into alcohol and carbon dioxide.
Proteolytic: proteases and *peptidases* break down the huge protein molecule into amino acids.
'*Maturalytic*': *oxidases* help trace-oxidation, and can lead to the appearance of brown oxidation products.

These are, of course, only a few examples of the multitude of enzymes present in yeast or any other cell.

Diffusion and Osmosis

Apart from a few odd exceptions (e.g. lactase) yeast enzymes are entirely intracellular and function only within the confines of the cell membrane, except, of course, when they have been liberated by autolysis. The cell-membrane is semi-permeable, and all substances needed by the cell enter in solution by *diffusion* through the outer wall and the cytoplasm. Only substances in solution can diffuse into or out of the cell, and this applies to oxygen and carbon dioxide as well as solids. The solute molecules are in continuous motion, and this determines their ultimate uniform dispersal and *diffusion* throughout the solution.

Simple sugars (e.g. glucose, fructose, sucrose, maltose), amino acids and dipeptides can all diffuse through the cell membrane, but large molecules (polysaccharides, proteins) are unable to do so. The latter need breaking down by extracellular enzymes derived either from autolysis (self-breakdown) of dead yeast cells, or from other external sources, before they can pass into the cell.

It is obvious that restriction of nutrient means a reduction in the growth of yeast cells, but the same situation can apply when there is an excess of food. Yeast reproduction ceases if the external sugar concentration is too great, a fact reflected by the use of syrup as a preserving medium. An understanding of this phenomenon involves a knowledge of the law of *osmosis*.

When two solutions of differing concentration are separated by a semi-permeable membrane, such as that of a yeast-cell wall, water passes through the membrane rapidly from the weaker to the stronger solution in an effort to equalise the differing strengths of the solutions. The pressure upon which this movement depends is called *osmotic pressure*, and it is by this means that plants draw their water from the soil. Just as over-manuring with soluble artificial fertilisers can form too strong a soil solution so that water is sucked out of plants, so if excessive amounts of sugar are in solution in the must, the osmotic pressure of the external solution will be greater than that of the cell contents. The cells accordingly lose water (which represents about 65% of their weight), the cytoplasm shrinks to the centre of the cell (*plasmolysis*) and metabolic activity ceases. This is the explanation of 'sticking fermentations' brought about by an over-sweet must. The sensitivity of wine yeasts to osmotic pressure, or *osmosensitivity*, varies according to the strain of the yeast and to the environment in which it has been propagated. Apart from ellipsoidal wine yeasts, certain species exist with low osmosensitivity characteristics, and examples of such osmophillic species as *Saccharomyces mellis* and *S. rouxii* will be mentioned in chapter 3.

Chapter 3 Yeast Classification

A summary of plant classification is set out:

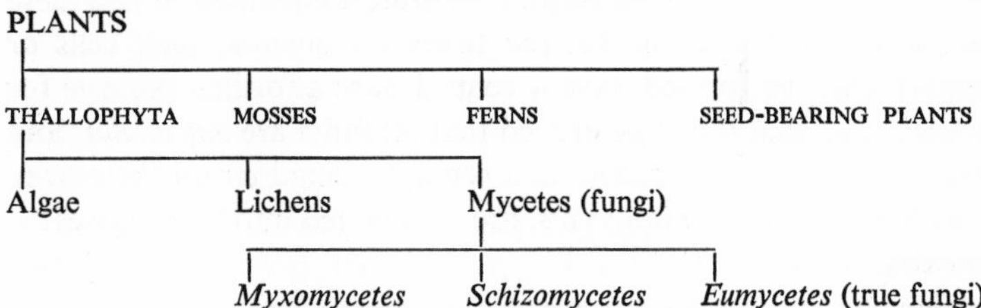

THALLOPHYTA (Greek *thallos* = young shoot, *phyton* = plant). Apart from certain unicellular forms, these are plants in which the constituent cells are arranged in filaments or form a more elaborate well-defined structure known as a *thallus*. Only in the higher algae is there any differentiation of the thallus into regions superficially comparable with the leaves, stem and roots of higher plants.

Mycetes (Greek *myketes*, plural of *mykes* = fungus). The fungi are distinguished from the algae, or seaweeds, by their lack of the green pigment *chlorophyll*, and are consequently unable to build up cellular material from carbon dioxide and water. Lichens are compound organisms of algae and fungi.

Myxomycetes (Greek *myxa* = slime, *myketes* = fungi). The origin of the name describes the appearance of these organisms, which may be variously regarded as primitive fungi or primitive *protozoa* (Greek *zoon* = animal).

Schizomycetes (Greek *schizein* = to cleave). These are bacteria, more usually considered as a completely independent group devoid of fungal affinities. Their name arises from their method of reproduction by transverse splitting. *Acetobacter* and *Lactobacillus* belong here.

Eumycetes (Greek *eu* = well, *mycketes* = fungi). It is to this class of true fungi that yeasts belong.

Before we leave this general classification in order to consider the taxonomic relationships of yeasts in greater detail, it might be mentioned that there is actually a theory that fungi represent a third kingdom on their own, co-equivalent with Plant and Animal kingdoms, and are not a subdivision of Plants at all. Whereas green plants are able to utilise carbon dioxide as the sole source of carbon (*photo-synthesis*), fungi resemble animals in requiring more complex food materials. On the other hand, many fungi obtain nitrogen in the form of inorganic salts in the same way as plants, while animals are able to utilise only organic sources such as proteins and amino acids. It has been observed by M. Langeron that fungi differ in that they never form a tissue, but are composed of tubes (*hyphae*), which, as a branching, intertwining mass, constitute a *mycelium*. When these tubes have cross-sections, i.e. the tubes are *septate*, such cells or compartments so formed have a central pore affording passage for cytoplasm, so that it can be argued that all fungi are *unicellular*, and hence entitled to be regarded as a separate kingdom on their own.

The 'true fungi', or *Eumycetes*, are subdivided into four classes:

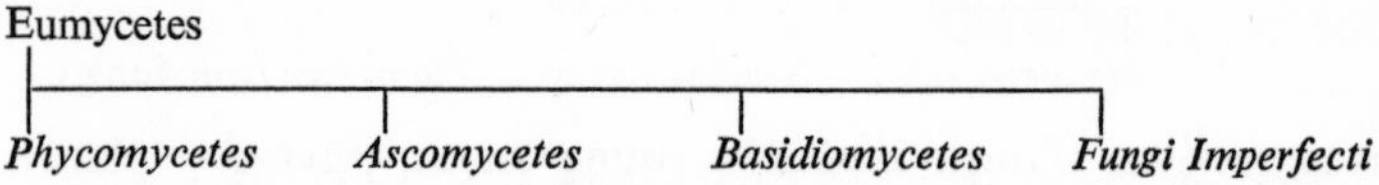

Eumycetes

Phycomycetes *Ascomycetes* *Basidiomycetes* *Fungi Imperfecti*

Phycomycetes (Greek *phykos*, seaweed). These alga-like fungi have 'non-septate mycelia', which means that the hyphae do not have any cross-walls, or *septa* dividing them into compartments, as do the other three classes. Various domestic moulds belong here.

Ascomycetes (Greek *ascus*, skin bag). The sexual spores of these fungi are produced in thin-walled cells called *asci*. There are some 15,000 species in this class, including the edible fungi called truffles, in which the asci are produced underground in a fruit-body. For us it has especial importance, because it contains the beer and wine yeasts.

Basidiomycetes (Greek *basidium*, club). To this class belong toadstools and mushrooms, and all those fungi which bear their spores on short projections called *sterigmata* at the ends of club-shaped cells known as *basidia*.

Fungi Imperfecti. This name is given to those fungi with an incomplete, or at least incompletely known, life-cycle, such as *Penicillium* and *Aspergillus* moulds. It is highly probable that many are really Ascomycetes which produce ascospores only under certain conditions or have perhaps no longer the power of doing so. They reproduce by means of *conidia* (Greek *konis*, dust; *eidos*, like), the term applied to asexual spores, or by fragmentation of the mycelium.

Taxonomy of yeasts

Now that we have at last tracked down wine and beer yeasts in the largest fungal group, we can proceed with their classification.

There is more general agreement on this than in the past, and their characteristic features have been settled as a means of separating them into genera and species. Even so, there is still some confusion as scientists put forward reasons for changing yeasts from one category to another, or for the formation of a new branch from an old one, so that the amateur has every reason for being baffled and bewildered. Matters are not helped when one writer refers to a yeast by an early name and another uses a more modern synonym for the same thing. Fortunately, the few yeasts of interest to the winemaker are reasonably clearly defined, and abstruse problems of borderline cases need not bother him.

Yeasts, a term which has no real scientific significance, are divided into three families:

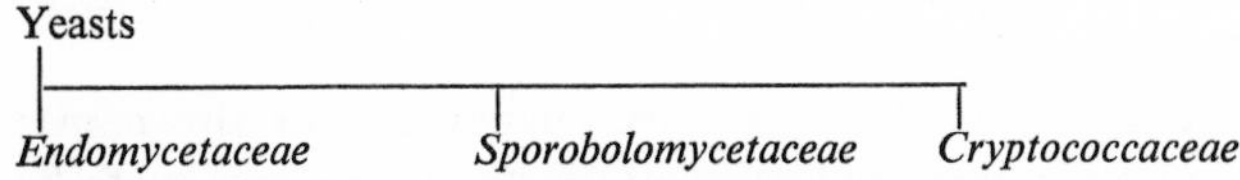

These names really convey quite simple meanings; family names always end *-aceae*, pronounced -áy-se-ee.

Endomycetaceae (Greek *endon*, within). The name refers to the spores within the ascus. These are the so-called 'true yeasts', because they produce ascospores and are true representatives of the Ascomycetes.

Sporobolomycetaceae (Greek *bolis*, missile). This very small family, with only two genera, is difficult to classify, because their spores are formed on small projections and are ejaculated into the air by a propulsive mechanism; they could well be primitive or reduced Basidiomycetes.

Cryptococcaceae (Greek *kryptos*, hidden; *kokkus*, seed). This family produces neither ascospores nor ballistospores, and the members are therefore examples of the Fungi Imperfecti. Either they have lost their power of producing sexual spores or do so by methods still uncertain.

Unlike moulds, yeasts are not usually identified by morphological characteristics alone, but also by biochemical tests. In other words, the shape and size of the cells, their methods of propagation and their modes of spore production, provide insufficient criteria, and additional chemical tests are applied. These methods include their ability to ferment different sugars and to utilise nitrate as the sole source of nitrogen, and sometimes their ability to derive their carbon from alcohol.

The Dutch School

In 1931 N. M. Stelling-Dekker published a monograph on spore-forming yeasts, and in 1934 and 1942 N. J. W. Kreger-van Rij did the same for those that do not produce spores of any kind. These two mycologists were working in the Dutch Centraalbureau voor Schimmelcultures, at the Laboratory of Microbiology of the Technical University at Delft. In 1952 J. Lodder and N. J. W. Kreger-van Rij co-operated in bringing out a single condensed volume that brought these previous works up to date. It is a most authoritative work, and although their nomenclature may be confusing to those accustomed to terms in earlier use, and not all of their observations may be generally acceptable to other mycologists, nevertheless their work brings order to a state of affairs where multiplication of synonyms was leading to chaos. Their nomenclature is used throughout this book.

While there is no need to concern ourselves with the names allotted to the sub-families and tribes of yeasts, it is interesting to be able to compare the relationship of those that are of practical value to the winemaker, and therefore a simplified table is given below.

Yeasts are divided into three families; of these we need consider only two: those that form ascospores, called *Ascosporogenous Yeasts*, and those that form no spores of any kind, called *Asporogenous Yeasts*, the prefix here meaning 'without'.

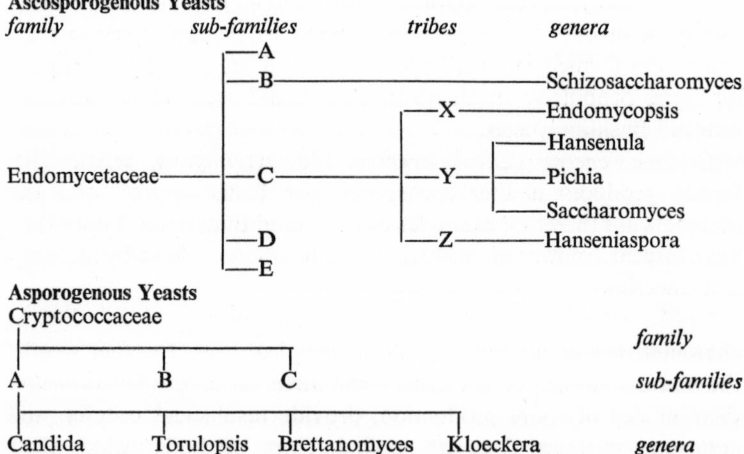

The genera mentioned in the tables can now be considered in detail. In referring to a particular yeast, it is usual to employ two Latin words, a method called 'binomial nomenclature'. The first word is the name of the *genus* (plural = *genera*), and the second is its *species*, a further subdivision of the genus. Thus we speak of *Saccharomyces cerevisiae* and *S. carlsbergensis*; *Saccharomyces* is the genus, which can be represented by the initial letter where it is clearly understood, and *cerevisiae* and *carlsbergensis* are two species of this genus.

Yeast Genera

1. Schizosaccharomyces

The distinctive feature of this genus, which gets its name from the Greek *schizo* = to split, is its reproduction not by budding but by fission. The cell grows lengthways, develops a partition wall and then splits along it. The name was first used in 1893 by Lindner, who isolated it from African millet beer, giving it the species name *pombe*, the local term for this drink. The natives practise a primitive form of malting, and since a starch-hydrolysing mould (*Aspergillus flavus*) is also present, most of the starch is converted to glucose before or during the fermentation. The other two species are *S. versatilis* and *S. octosporus*, the latter characterised by the formation of eight spores instead of the usual four. As sporulation follows immediately after conjugation with this genus, the cells are haploid.

2. Endomycopsis

The species *Endomycopsis fibuliger* deserves a place because it is often mentioned in connection with the fermentation of rice wines or other beverages of a farinaceous nature. The distinctive feature of the genus is the production of extra-cellular *amylase*, that is, an enzyme which is able to break down starch into fermentable sugars, although it ferments these very weakly. As one might expect, these yeasts are often isolated from flour, a natural source of food for them. There are six species and one variety.

3. Saccharomyces

This is the genus that has provided yeasts for baker, brewer and vigneron, and is consequently of the greatest industrial importance.

The name originates from the Greek words *sakchar* = sugar, and *mykes* = fungus, referring to the strong sugar-fermenting properties of the genus in general. There are thirty species, and as this yeast is of such importance all the names are set out overleaf:

S. cerevisiae
S. cerevisiae var. *ellipsoideus*
S. pastorianus
S. rouxii
S. rouxii var. *polymorphus*
S. exiguus
S. marxianus
S. bailii
S. logos
S. bayanus
S. willianus
S. uvarum
S. delbrueckii
S. delbrueckii var. *mongolicus*
S. carlsbergensis
S. fragilis
S. lactis
S. rosei
S. chevalieri
S. bisporus
S. pastori
S. fermentati
S. heterogenicus
S. microellipsi
S. oviformis
S. mellis
S. italicus
S. florentinus
S. acidifaciens
S. steineri
S. fructuum
S. elegans
S. veronae

For a long time all spore-producing yeasts, and even some that did not form spores, were lumped together as various species of the *Saccharomyces* genus. Then Hansen, in 1904, and others after him, gradually restricted the genus and separated certain species to form new genera such as *Pichia* and *Saccharomycodes*. Among changes made by the Dutch school is the merging of the genus *Zygosaccharomyces*, first created as such in 1901, and its species with the genus *Saccharomyces*. The main distinction between the two genera is one of genetics: *Zygosaccharomyces* yeasts are characterised by the occurrence of sporulation immediately after the conjugation has taken place. The ascus is thus the only diploid phase of their life-cycle, and the ordinary cells are haploid. *Saccharomyces*, on the other hand, normally bud vegetatively after conjugation for some time before sporulation takes place, so that they are predominantly diploid. The modern tendency is to regard these diploid and haploid phases as controlled and directed by conditions not yet fully comprehended, so that the same organism emphasises one or the other phase according to different circumstances. Another example of this phenomenon is the *Torula*, or more correctly *Torulopsis*, yeast, which is a type of *Saccharomyces* that has become *permanently* haploid, and has lost its power of sporulation.

On such grounds, *Zygosaccharomyces* spp. are regarded as the prolonged haploid phase of the same organism that in its diploid phase was previously distinguished as *Saccharomyces*, and therefore the term *Zygosaccharomyces*, which is still often used in articles on yeast, is regarded merely as a synonym for the genus *Saccharomyces*, and not another genus in its own right.

Another genus, *Torulaspora*, created in 1904 and deriving its name from the fact that it looks like a *Torula* cell but forms spores, has also been rejected as a separate genus and its spp. attached to *Saccharomyces*.

In addition to the merging of these two genera, *Zygosaccharomyces* and *Torulaspora*, in *Saccharomyces*, large numbers of synonyms have been listed as redundant on the grounds that the species so named are identical with the thirty accepted species of this genus. Thus *S. ilicis*, *S. vordermannii*, *S. cratericus*, *S. anamensis* and many more are merged in the species *S. cerevisiae. Zygosaccharomyces nadsonii*, *Z. major*, *Z. rugosus* and others are merged in the species *S. rouxii* (a yeast found in concentrated sugar solutions, and named after a mycologist, Roux), and so on.

A. Beer Yeasts

In 1838 J. Meyen first gave the name *Saccharomyces cerevisiae* to brewers' yeast. The term *cerevisiae* is probably a corruption of *cervisiae*, a Latin word of Celtic origin, meaning 'of beer'. The goddess of agriculture is named Ceres, and confusion between the two is easy. Hansen restricted the use of the name to a typical English beer yeast that forms on the top of the wort. It may be regarded as the perfect form of *Candida robusta*. Later he named another species, a typical continental bottom-fermenting lager yeast, *Saccharomyces carlsbergensis*. The specific-name is from the well-known Danish brewery, and means 'of Carlsberg'.

Emil Christian Hansen, the first Director of the Carlsberg Laboratory in Copenhagen, pioneered the isolation of pure strains of cultivated yeast, and initiated the use of pure cell cultures in starting fermentation in sterile must. He diluted the yeast in wort gelatine so that there was one cell in every two drops. Then, by means of a 'moist-chamber slide', he followed the development of a selected single cell by a microscope until it had become a colony of cells ready for transference to a sterile medium.

The different behaviour of top and bottom yeast has given rise to two quite distinct brewing techniques. Top fermentation, or ale process, predominates in Britain and is widespread in Belgium, whereas bottom fermentation, or lager process, is generally employed in continental Europe and America. Distinct differences in the resulting character of the beers so brewed are due not only to biochemical differences between the yeasts but also to differences in the brewing procedures adopted to suit them, and these are discussed in chapter 20.

It is generally believed that the parent type of all culture yeasts was the English top yeast, a telling argument for this view being that bottom yeasts generally contain a proportion of top yeast cells and can be easily converted into top yeasts. In that case, it would seem that bottom yeasts have been developed from them by genetical segregation. Then, too, the top yeasts are more robust in character and stand up to adverse conditions more favourably, as well as producing spores more easily. Both types have been cultivated for centuries by brewers; bottom yeasts are considered to be relatively more recent in origin, and top yeasts to be a 'wilder' strain.

There is little difference of density between cells of the two types, and the reason for their behaviour in the wort seems to be due to the fact that top yeasts tend to become attached in clusters so that the rising gas carries them to the surface. In the early stages, too, the buds do not separate immediately from the parent cell, so that chains of cells are formed, thus increasing the tendency to trap bubbles of carbon dioxide. A further distinction between the top and bottom types is the preference of the latter for a lower optimum fermenting temperature, a point that is an important consideration in the brewing procedure. Finally, they vary in their enzymic constitution, and this forms a useful test in separating them. Both yeasts are able to break down the sugar *raffinose* into fructose and melibiose, but only the bottom yeast can hydrolyse the latter sugar into glucose and galactose, and thus ferment raffinose completely.

S. carlsbergensis	raffinose		fructose	All three
Bottom yeast		melibiose	glucose	sugars
			galactose	fermented
S. cerevisiae	raffinose		fructose	fermented
Top yeast			melibiose	unfermented

This result is due to the fact that top yeast contain the enzyme *invertase*, which can split fructose from raffinose, but not the enzyme *melibiase* contained by the bottom yeast, which is needed to hydrolyse melibiose into glucose and galactose.

B. Bread Yeasts

A plausible suggestion to explain the origin of yeast or 'leaven' to aerate bread is that at some time in the distant past a piece of unbaked dough was left exposed to the air, with the result that airborne yeasts settled on it and started to ferment. When it was baked, the resultant improvement, compared with unleavened bread, led to repetition of the process until someone with more awareness than the rest reserved a piece of the leavened dough as a 'starter', thereby removing the uncertainty of spontaneous leavening. Until the middle of the nineteenth century little change took place, various 'barms' or yeasts from the brewery being used as the agent of panary fermentation. Then the increasing industrialisation of towns encouraged the commercial production of bread by large bakeries, and the home baking of bread dwindled away. The result was that a demand for better bread yeast arose on a large scale, and ways of providing it started the new industry of compressed yeast. In 1860 the 'Vienna process' was producing improved qualities of yeast by the aeration of fermenting malted grain infusion, and Louis Pasteur in the same year in his *Notes on Alcoholic Fermentation* was drawing attention to the increased reproduction of yeast cells under aerobic conditions. Advances were concurrently made on the engineering side in supplying the machinery needed by the factories, and in 1868 the new industry of compressed yeast manufacture was inaugurated by Fleischmann in America.

Although improvements were being made at the same time in flour milling and baking techniques, it was fundamentally the introduction of compressed yeast that produced the revolution in the baking industry during the last hundred years. Thanks to the researches made in yeast technology, the fermentation period for dough that lasted anything from twelve to sixteen hours with the old barm or brewers' yeast has been reduced to the modern 'short process' of about two hours.

The yeasts used for brewing and baking are markedly different, and they are not readily interchangeable. Brewery yeast is dark—most bakers have a preference for a light-coloured yeast, although the small 2% used in bread is unlikely to darken the product—and it is rather bitter because of material derived from the hops. Further, it works too slowly at bakery temperatures, so that the loaf would resemble the domestic type being made at home from barm at the beginning of last century: close textured and of small volume, of very strong flavour and tending to be bitter, and perhaps dark in colour. Distillery yeast is more suitable for baking, although why it ferments flour more rapidly than a brewery yeast when the action of both on a sugar solution is about the same, is difficult to explain. At any rate, no baker would use brewery yeast in his bread and hope to get away with it today.

On the other hand, bakery yeast is unsuited for beermaking. It is true that alcohol is produced during panary fermentation; although it is largely lost in the baking process, about 0·5% remains to add to the flavour. The baker, however, is not so concerned with this. His aim is the opposite of the manufacture of potable drinks; he wants the minimum of alcohol and the maximum of gas for his aeration purposes, and the methods of compressed yeast manufacture encourage this strain of yeast. High temperature, strongly aerated conditions and 'exponential feeding', i.e. the addition of sugar and nutrients to keep pace with cell development, all go to influence the character of the yeast.

Yeasts are modified by the environment in which they are propagated; it is known, for example, that a yeast that has been produced in a cereal grain wort is much richer in *maltase* than a yeast propagated exclusively in a wort made from molasses. Thus it is quite possible that bakery yeasts, if grown under brewery conditions for a number of generations, might take on the character of brewery yeasts, and vice versa; environment is virtually as important as heredity in determining the character of a yeast species. Therefore, although brewery and bakery yeasts are both strains of the same species of the same genus, *Saccharomyces cerevisiae*, they are markedly changed by their environmental upbringing with regard to their enzymic make-up. To adapt a proverb, a yeast is known by the enzymes it keeps, and if the difference in growth conditions do not change the quality of enzymes present, at least they markedly affect the quantity. Bakery yeast contains a system of enzymes conducive to the oxidation of sugar to carbon dioxide and water; brewery yeasts contain an enzyme system suited to the conversion of sugar to alcohol. These differences are fundamental to their respective employment.

C. Wine Yeasts

The point just made also applies to wine yeasts. These again are members of the great *S. cerevisiae* group, although Hansen regarded them as a separate species. They are more ellipsoid in shape than the round or ovate cells of brewery and bakery yeasts, with a length to breadth ratio of 2 : 1; 'long oval' is a good description of their appearance. Hansen decided that their differences were sufficient to justify specific status, and he restricted the name *S. ellipsoideus* to them. This name had already been coined by Reess in 1870, and loosely applied to yeasts that fermented grape and fruit juice. In the nomenclature of the Dutch school, these yeasts are classified as a variety of *S. cerevisiae*, and consequently named *Saccharomyces cerevisiae* var. *ellipsoideus*. In general articles, however, one will see them briefly described as 'ellipsoidal yeasts', or 'true wine yeasts'.

Amateur winemakers in the early stages of their hobby often make use of granulated bread yeasts, and there are points in favour of this practice for beginners. The inoculation of the must by sprinkling the yeast on the surface dispenses with the use of the 'starter-bottle' needed for wine-yeast cultures, thus simplifying the process. Further, it is known that bakery yeasts that have been cultured by the exponential systems prevailing in compressed-yeast factories have a diminished *osmosensitivity*, i.e. they are more sugar tolerant, and as beginners commonly tend to start their fermentation with an excess of sugar in the must, this is an argument in favour of their use. Progressive winemakers, however, inevitably graduate to the true wine yeast because, as has been discussed above, only a yeast that has been propagated under a vinous environment can elaborate the developed glycolytic system of enzymes able to provide the winemaker with the type of quality wine he wishes to make.

In general, a good wine yeast should have the following four properties:

1. High alcohol tolerance, i.e. the yeast should continue to ferment despite the increasing concentration of alcohol, giving stronger, drier wines with up to 16% alcohol v/v, or even up to 18% v/v where the yeast is 'fed' by periodic additions of sugar in small amounts.
2. Good degree of agglutination, i.e. the tendency of the yeast to flocculate into small lumps that give a cohesive sediment as fermentation ceases, so that racking is simple and the wine clears easily. On the other hand, a powdery yeast that tends to remain in suspension is often to be associated with a high-alcohol producing yeast, and a compromise may be necessary.
3. Steady, persistent fermentation capacity; this leads to wines of better quality than when the fermentation falls away after a tempestuous start.
4. Absence of unpleasant flavours generated by dead and dying cells. Wine left on bakery yeast for any length of time takes on a musty, new-bread flavour.

A large number of strains of *S. ellipsoideus* are now available to the amateur, usually taking their name after the district of their origin, such as Burgundy, Bordeaux, Rheingau and so on. A view held by many is that the subtle variations of flavour and bouquet that exist between the same type of wine from different vineyards are due not only to differences in the soil, to the positioning of the vines and to the variety of the grape but also to the different characters of the various strains of ellipsoidal yeast. At first it was thought that differences in the yeast population were related to the grape variety, each variety carrying a particular strain, or mixture of strains on its skin, but it is now believed that the district of origin, in a topographical sense, is the controlling factor. If this is true, then the use of a selected wine yeast from the Burgundy district in a suitable wine-must should confer on it something of the traditional Burgundian quality. So many factors enter into this problem that it is difficult to pronounce decisively on it, and it remains a question of individual opinion. Reference should be made to 'Environmental Modification'.

As an indication that distinct differences *can* exist between strains of wine yeast from different wine districts, one can quote the Sauternes yeast from the district of that name in the Bordeaux area. This yeast ferments glucose slowly and reluctantly, but produces alcohol readily from fructose. Its action on grape juice, which contains both sugars, gives a wine which, although having the alcoholic content of a normal table wine, is decidedly sweeter than usual because of the unfermented glucose. Such wines are otherwise obtainable, as in sherry production, only if the fermentation is artificially arrested, by the addition of alcohol, for instance. Presumably the yeast cells reject the unwanted sugar by having a selective cell membrane, so that one sugar is able to penetrate easily and be fermented, while the other remains outside the cell and sweetens the wine. Other possibilities are that this yeast has a fructokinase but no hexokinase, or that it has a specially powerful glucose-6-phosphatase. Further examples of wine yeasts with characteristic action are given by the champagne yeasts, which are relatively insensitive to the presence of carbon dioxide and remain active even at high gas concentrations, and the sherry yeasts with their 'flor' formation.

D. Sherry Yeasts

A species of *Saccharomyces*, named *S. fermentati*, is as the name suggests a strong fermenter of sugars, with a high alcohol tolerance and the tendency to form little spots of 'pellicle' or skin on the surface of the must. A strain of this yeast, sometimes regarded as another species and named *Saccharomyces beticus*, from the old Roman name for the province of Andulasia, is regarded as the typical sherry yeast, as used in Spain and California. It tolerates the high alcohol content of sherry, gives the typical rancio flavour and taste to these wines, and forms a sherry 'flor' or skin on the surface of the must. Usually, as mentioned later, other yeast species and genera are present as well.

The formation of a pellicle entirely precludes the access of oxygen to the wine below; the consequent absence of dark oxidation products gives a wine, termed a 'fino', having a distinctly pale colour and contrasting sharply with the 'oloroso' produced in the absence of a pellicle. Whether or not a flor is established is at present largely a matter of chance, and the factors influencing its formation are imperfectly understood.

Other Species of Saccharomyces

1. *S. pastorianus.* This is a spoilage yeast that gives a disagreeable smell and flavour to beer. It is named after Pasteur, who studied its effect in his 'Études sur la Bière'. It ferments glucose, sucrose and maltose, but not galactose, and is aptly described as 'sausage-shaped'.

2. *S. rouxii.* Forms of this yeast are often found in honey, jams and syrups, since they are *osmophilic*, that is, they can ferment high concentrations of sugar. They ferment glucose and maltose, the latter very weakly, but not sucrose. Their specific name is derived from Emile Roux, who first drew attention to the fact that not all yeasts that ferment glucose can also ferment sucrose. In this species, as ascus formation follows quickly after conjugation, so the haploid phase tends to predominate; several *Zygosaccharomyces* synonyms are still in use.

3. *S. fragilis.* Only a limited number of yeasts can ferment milk sugar (lactose) because the majority lack the enzyme *lactase* needed to hydrolyse this sugar into glucose and galactose as a prelude to fermentation. *S. fragilis,* so-called because of the fragility of its cells, is a typical lactose-fermenting yeast. In addition to lactose, glucose and galactose, it also utilises sucrose, but is unable to ferment maltose. It is the perfect form of *Candida pseudotropicalis.*

Lactose-fermenting yeasts are of value, first, in the production of fermented milk drinks, and secondly, in their use as food yeasts, because they will grow on milk waste and whey. *Koumiss* is a staple article of diet in the Caucasus, and consists of fermented mares' milk or occasionally camels' milk. Elsewhere in Russia, another fermented drink, *Kefir,* is derived from cows' milk, and similar products are prepared in Denmark (*Koeldermilk*) and Egypt (*Leben*). A number of different yeasts usually take part in the fermentation together with Lactobaccili; whereas the former convert lactose to alcohol, the latter produce lactic acid. Koumiss has an alcohol content of 1–2% together with 0·5–1·5% lactic acid, 2–4% lactose and 1–2% fat. *Kefir grains,* used in the production of kefir, look like popcorns and consist of a symbiotic mass of the appropriate micro-organisms.

4. *S. lactis.* As the name implies, this is another lactose-fermenting

yeast. It represents the perfect form of *Torulopsis sphaerica*, and is associated with *S. fragilis* in the production of Koumiss.

5. *S. mellis*. The specific name means 'of honey', and as is the case with *S. rouxii*, it frequently appears in concentrated sugar solutions. It is only capable of fermenting glucose. Synonyms under the generic name *Zygosaccharomyces* are still often employed.

4. Pichia

Hansen first isolated this yeast from 'exudation of elms' near his Carlsberg laboratory. He originally described it as *Saccharomyces membranaefaciens*, but sixteen years later (1904) transferred it to a new genus, *Pichia*, named after the Italian mycologist Pichi.

The characteristic feature of this genus is its formation of a pellicle, a dry, wrinkled and creamy-coloured skin, on the surface of the must. The various species prefer aerobic conditions and produce little or no alcohol, although progressive adaptation to a fermentative metabolism may occur in the absence of air. In so far as oxygen is available, they oxidise available substrates, including sugars and the alcohol produced by other yeasts, to organic acids and ultimately to carbon dioxide and water; as a consequence of the acid production, their presence leads to the appearance of various esters, especially ethyl acetate.

Hohl and Cruess in 1939 examined the flor of different fino sherries and found this genus to be present, in association with *Saccharomyces* species. Although this implies some aerobic breakdown of the alcohol, the resultant ester formation contributes to the rancio flavour of fino and amontillado sherry. *Pichia* yeasts are unwanted contaminants in table wines, but their activity is curtailed under the anaerobic conditions which result when an air-lock is fitted.

Of the four species which constitute the genus, the type species, *P. membranaefaciens*, achieves at best a very feeble alcohol production; it is probably the perfect form of *Candida mycoderma*. *P. fermentans*, however, does produce a reasonable conversion of glucose to alcohol, and does not promote ester formation.

Pichia is very closely related to the genus *Hansenula*, and some authorities have felt that the two should be merged into one genus. However, Lodder and Kreger-van Rij have kept them as separate genera, mainly on the grounds that *Hansenula* produces a vigorous fermentation and is able to utilise nitrate as a source of nitrogen. Since relatively few yeasts are capable of reducing nitrate, the ability to do so represents a valid taxonomic criterion.

5. Hansenula

Hansen was also responsible for the original description of this genus, which he isolated from Bavarian brewers' yeast. Originally named *Saccharomyces anomalus*, it was later (1904) transferred by Hansen to a new genus, *Willia*, but the generic name was changed yet again to *Hansenula* (1919) because of confusion due to the relegation of other yeasts to *Willia*.

Hansenula is a pellicle-forming yeast and occurs, together with *Pichia*, in sherry flor. Although capable of vigorous fermentation, it is also often responsible for pronounced ester formation and in the absence of sugars can utilise alcohol as the sole carbon source. The genus contains nine species. The type species, *H. anomala*, ferments sucrose as readily as glucose, and can also utilise galactose and maltose, although rather less effectively. It is usually accepted as the perfect stage of *Candida pelliculosa*.

6. Hanseniaspora

This name, again commemorative of E. C. Hansen, is used for a genus of spore-forming apiculate yeasts. 'Apiculate' means 'possessing a short, sharp point', and the cells of such yeasts are either pointed at both ends like a lemon or pointed at one end like a pear.

Apiculates, which are usually present on fruit and in the air, are the most troublesome of 'wild yeasts' in so far as winemaking is concerned. If spontaneous fermentation is allowed to take place without sterilisation of the must they are the first of the various yeast genera to start fermenting, for a rapid start is a characteristic feature. Apiculate yeasts are, however, especially sensitive to alcohol, so that having produced an alcohol concentration of about 4%, and this is possible in 36 hours, they become progressively inactive. Unfortunately, the flavour and quality of the wine may have already suffered. In a spontaneous fermentation inactivation of the apiculates is followed by the appearance of *Saccharomyces* spp. as the main fermenting agents. In wines made from fruit other than grapes, there is no certainty that wine yeasts will be present and able to continue the fermentation, in which case the wine 'sticks'; the production of alcohol stops before the amount of sugar present has been converted to the desired extent, and one is left with a weak, tart and over-sweet cordial. For reasons such as these, it is nowadays usual to sterilise the must before inoculating it with a selected yeast culture.

The amount of acid formed by apiculates (0·5–1·2 g/l.) is about double that produced by ordinary wine yeasts, and gives the wine a sharp cider-like flavour. For this very reason, some manufacturers tolerate the presence of apiculates in the production of draught cider; since the yeasts are unable to ferment sucrose, they are especially acceptable in the production of sweet cider because any sucrose added to the must remains unchanged.

Apiculate yeasts were first described by Reess in 1870, under the name of *Saccharomyces apiculatus*. Later, when it was found that some formed spores and others did not, the sporogenous forms were referred to a new genus, *Hanseniaspora* (Zikes, 1911). Sporulation with apiculates is always uncertain, their ability to do so disappearing when they have been cultivated for some time, so that there has been considerable confusion with certain strains whether they are sporogenous or not, and what number and shape of spores they produce. Amalgamation of the two species, *H. guilliermondii* and *H. melligeri*, under the name *H. valbyensis* leaves this as the sole species of the genus. It was isolated by Klöcker in 1912 from the soil of a district near Copenhagen called Valby, hence its specific name. It is accepted as the perfect stage of *Kloeckera apiculata*. Since the name *S. apiculatus* was originally given to an asporogenous apiculate yeast, it is not applicable to the sporogenous forms.

7. Candida

So far we have been dealing with genera of sporogenous yeasts; with *Candida* we come to the asporogenous yeasts, sometimes called 'imperfect yeasts' or 'pseudo yeasts'. It has already been seen that many of these are the imperfect forms of spore-producing yeasts, their ability to form spores having been lost.

Candida is a large genus, with thirty species. A very large number of synonyms occur, increased by the fact that even the generic name has varied in the past between *Monilia*, *Oidium*, *Syringospora*, *Mycotorula* and others. The term *Candida*, from the Latin *candidus* meaning 'white' and referring to the whitish appearance of the colonies, was not created until 1923 by Berkhout. A number of species are of particular interest:

Candida albicans, the type species, is an animal pathogen. Its unchecked multiplication in the membranes of the mouth and throat, especially of children, leads to the disease known as 'thrush'.

C. mycoderma forms a creeping pellicle, yellowish-white in colour, on the surface of a must. Under aerobic conditions it spreads and oxidises available substrates, including alcohol, to carbon dioxide and water, in effect converting the wine to soda water. In the absence of air, it is only able to utilise glucose, and its fermentative ability is slight. Unlike the two sherry flor yeasts, *Pichia membranaefaciens* and *Hansenula anomala*, it contributes nothing to the flavour and is a spoilage yeast only; it is rendered harmless by the use of an air-lock on the fermentation vessel. It multiplies rapidly in the small residue of table wine at the bottom of neglected bottles.

C. mycoderma is also known as *Mycoderma vini* (from the Greek *derma*=skin), but this is most unsatisfactory, because this name has been variously applied, at least as far back as 1822, to a number of yeasts, moulds and bacteria. In 1870 Reess included all the pellicle-forming yeasts in the term *Saccharomyces mycoderma*, and again one still sees from time to time this vague term in use. Country folk know the yeast as 'flowers of wine'; it is more probable that this usage is an alternative spelling for 'flour', the pulverulent form of any substance, rather than that it originates from a supposed resemblance to floral morphology.

C. pseudotropicalis is interesting, since it ferments lactose and galactose as well as glucose and sucrose, and consequently it is used in the fermentation of milk beverages such as koumiss and kefir. In Germany it has been grown on waste whey and used as a food yeast. The variety *C. pseudotropicalis* var. *lactosa*, which has larger cells, was originally isolated from buttermilk. The specific name refers to its similarity to *C. tropicalis*, a human pathogen capable of infecting the bronchii and which was first isolated from a patient suffering from bronchomycosis in Ceylon: hence the name.

C. lipolytica takes its name from the Greek *lipos*=fat and *lysis*= loosening, and contains a highly active *lipase*, an enzyme capable of hydrolysing a variety of the naturally occurring fatty substances collectively known as lipids. It can assimilate glucose, but does not produce alcohol. The original organism was isolated from stale margarine in 1921, and other forms have been found on olives.

C. reukaufii, originally described by Reukauf, is a most remarkable yeast which survives the winter inside bees' stomachs and is later transferred to flowers, where it can be found growing in the nectar. Occasionally cells of the colony are especially large and contain prominent oil droplets; they closely resemble the 'pulcherrimi' cells of another species *C. pulcherrima*, the specific name meaning 'very beautiful'. These yeasts have occasionally been exploited, for example, in Germany during the wars, as a source of fats.

C. utilis until recently was known as *Torulopsis utilis*, the specific name meaning 'useful', but it is now included with the *Candida* genus because it forms what is called a *pseudomycelium*, composed of branching chains of cells. Its value consists in its use as a 'food yeast'. Various types of yeast have been grown for this purpose, for not only do they represent a relatively concentrated source of protein and a rich supply of vitamins, but they can be grown quickly and in bulk on cheap carbohydrate material. During the Second World War *C. utilis* was grown in Germany on sugars derived from the acid hydrolysis of wood, and in Jamaica on molasses and surplus sugar solutions. The product, a yellowish brown powder containing about 45% protein, was marketed as an additive for drinks, soups and bread, but its pronounced yeasty flavour, although slighter than that of *Saccharomyces*, has resulted in its main application today as a constituent of cattle food. The widespread use of yeast protein as an alternative to protein from other sources has in any case been challenged: not only does the yeast product contain little methionine, an essential amino acid, but it is also a relatively rich source of purines, nitrogenous bases having diuretic properties and potentially capable of contributing to the formation of kidney stones.

Two mutants of *C. utilis* are recognised. The first (*thermophila*) multiplies well at tropical temperatures, and the other (*major*) is characterised by the large size of its cells.

8. Torulopsis

Yeasts included under the generic name *Torula*, as used by Pasteur and Hansen, were transferred (Berlese, 1894) to the new genus *Torulopsis* (Greek *opsis*=similar) on the grounds that the former name had also been extended to filamentous moulds. The Dutch school has further separated all non-fermenting, capsule-forming species under the name *Cryptococcus*, only the fermenting forms remaining in *Torulopsis*.

Torulopsis spp. are usually present in wild yeast populations and are frequently associated with moulds. Their fermentative activity gives an alcohol concentration of about 4%, but since they are capable of producing sliminess in the wine, they are regarded as spoilage yeasts. Their activity tends to be depressed if apiculates are present in large numbers.

The genus contains twenty-two species, probably representing *Saccharomyces* spp. which have lost their ability to produce asci. The type species is *T. colliculosa*, meaning 'with small elevations' and referring to tiny warts on the cell walls; it ferments glucose and sucrose and is considered to be the imperfect form of *Saccharomyces fermentati*. Some species are osmophilic and occur in concentrated sugar solutions: the origin of *T. lactis-condensi* is obvious. *T. versatilis* is a yeast which ferments both maltose and lactose, a relatively unusual accomplishment. Lindegren supports the view already expressed that *Torula* or *Torulopsis* yeasts represent *Saccharomyces* spp. that have at some time in the past become permanently haploid.

9. Brettanomyces

This organism was isolated from English beer in 1904, and is distinguished by the secondary fermentation that it produced. It was put into the *Torula* genus with the specific name *brettanomyces*, from the Latin *Brettano*, referring to its occurrence in British breweries. Later the name was used as a separate genus, with four species.

Beers made by spontaneous fermentation are of widespread occurrence in Africa and are also appreciated in Europe: the acid Lambic and Faro beers of Belgium are very popular, at least with the local population. The latter products result from the action of *B. lambicus* and *B. bruxellensis*, which are capable of fermenting glucose, sucrose and maltose. They also form pellicles, and under aerobic conditions will oxidise alcohol to acetic acid. This process is so pronounced that the yeasts tend to kill themselves; it is usual to neutralise the excess acidity by adding calcium carbonate to the wort.

Their fermentation of sugars under anaerobic conditions is a slow process, but the yeasts are characterised by a high alcohol tolerance. As a result, they are capable of starting a secondary fermentation in breweries and so producing trouble in finished beer. Sometimes, however, strains are allowed to cause a late fermentation in cask beers; they then convert to alcohol not only residual sugars but also some of the dextrins (starch breakdown products) present in the beer, thus resulting in the production of a stronger beer.

The species *B. anomalus* possesses the property, unusual in a beer yeast, of fermenting lactose but not maltose; hence its specific name.

10. Kloeckera

When Reess isolated apiculates, he named them *Saccharomyces apiculatus*, although he had not found any to sporulate. Klöcker objected to asporogenous yeasts being included in *Saccharomyces*, and he suggested the name *Pseudosaccharomyces apiculatus* for them, retaining *Hansenia* for those producing spores. Neither of these terms has survived, for the sporogenous apiculates are referred to *Hanseniaspora*, as seen above, and in 1923 Janke proposed the present name *Kloeckera* for the asporogenous forms.

Their effect on the must is very similar to *Hanseniaspora*, and the same comments apply to them. Glucose only is fermented. The type species is *Kloeckera apiculata*, and there are eight species in all.

Distinctive types of yeast

types	*names*	*comments*
beer	*Saccharomyces cerevisiae*	top fermenter, British ale.
	S. carlsbergensis	bottom fermenter, continental lager.
lambic beer	*Brettanomyces lambicus* *B. bruxellensis*	acid produced under aerobic conditions.
wine	*S. cerevisiae* var. *ellipsoideus*	high alcohol tolerance.
honey, syrup	*S. rouxii* *S. mellis*	osmophillic; predominantly haploid.
milk	*S. fragilis*	lactase.
	S. lactis	lactase.
	Candida pseudotropicalis	imperfect form of *S. lactis*.
fat	*Candida pulcherrima* *C. reukaufii*	fat producers.
	C. lipolytica	lipase.
starch	*Endomycopsis fibuliger*	amylase.
	Schizosaccharomyces pombe	found in millet beer.
pellicle	*Pichia* spp.	ester formation, poor fermenter.
	Hansenula spp.	ester formation, good fermenter.
	Candida mycoderma	spoilage yeast, imperfect form of *P. membranae faciens*.
spoilage	*Saccharomyces pastorianus*	spoils beer.
	Torulopsis spp.	imperfect forms of *Saccharomyces*
	Hanseniaspora spp.	spore-forming apiculates.
	Kloeckera spp.	asporogenous apiculates.
food	*Candida utilis*	rapid protein synthesis.
pathogenic	*Candida tropicalis*	bronchial infections.
	Candida albicans	throat infection thrush.

Wild Yeasts

This term is loosely but usefully employed in winemaking to indicate any yeast that is not wanted in the must but which may establish itself there unless steps are taken to prevent it. It must be remembered that true wine yeasts are found 'wild' on ripe grape skins, but always accompanied by less-desirable types together with wine-spoiling moulds and bacteria. Examination of wild yeast population can be summarised as follows:

Fruit

Californian grapes were found (Mark and McClung, 1940) to carry representatives of at least seven yeast genera on their skins. An impression of the relative importance of each genus is given by the following numbers, although these do not, of course, represent the actual cell number:

Saccharomyces (mostly *S. ellipsoideus*)	118
Candida	26
Torulopsis	25
Kloeckera	16
Hanseniaspora	11
Pichia	6
Hansenula	4

Lund (1954) estimated that between 10^3 and 10^7 yeast cells were present on strawberry, gooseberry and raspberry fruits, the greatest numbers being on decomposing specimens. One *Saccharomyces* species and many asporogenous types (*Candida, Torulopsis Kloeckera*) were present. Hansen, investigating the occurrence of *Kloeckera apiculata*, found it to be especially common on ripe strawberries, cherries, plums, gooseberries and grapes, but less frequent on red currants, raspberries and rowanberries.

Flowers

Flower nectar is a normal yeast habitat, where asporogenous forms are particularly well represented.

Vegetables

Lund also found yeasts, mainly *Candida* and *Torulopsis*, on damaged mangolds.

Spontaneous Fermentation

Spontaneous fermentation of non-sterile musts is initiated by the activity of the apiculates *Hanseniaspora* and *Kloeckera*, followed closely by *Candida* and also by *Torulopsis*, if the latter is not suppressed by the apiculates. After about two days *S. ellipsoideus* becomes predominant, but the wine may already have been spoilt. It is also possible that development of the *Saccharomyces* fermentation may be delayed, or the yeast may not be present, so that the fermentation 'sticks'. A further factor is the degree of contamination with moulds and bacteria.

It is true that many of the smaller continental vignerons still prefer the traditional spontaneous fermentation for their wine in place of the modern method of inoculating a sterile must with a vigorously fermenting selected yeast culture. It is true, too, that the mixture of yeasts and acid-resistant bacteria that enter into such a fermentation can produce a unique wine when conditions are favourable. It must be remembered, however, that in such a wine-producing district grapes have been grown for centuries, and some sort of consistency in the balance of micro-organisms on the grapes has been successfully proved, although natural conditions can always vary to the extent of upsetting this association, and spoilage can result. To deduce from this practice that the yeasts and bacteria on fruit of unknown origin fermented in a kitchen are likely to prove equally effective in the majority of cases for fermentative purposes is a fallacy. Even where the employment of selected cultures is not observed abroad, there is a growing practice of sulphiting the must in order to reduce the competition from wild yeasts present and encourage fermentation by the more sulphite-resistant wine yeasts.

Chapter 4 Yeast Nutrition

Although there is considerable variation in the way in which plants and animals obtain their food requirements, they are all similar generally in the main reasons why they need food. Nutrition is required by living organisms for two main purposes:

1. To supply energy for the activity of living.
2. To supply material for the building and repairing of cells.

These two purposes of nutrition will be taken separately, as they are served mainly by two different types of food substances. The best approach to an understanding of these is via a simple basic knowledge of the respiration and digestion of animals and humans, and we shall deal first with the supply of energy.

A. Carbohydrates: Combustion and Respiration

For complete combustion, oxygen must be available. We all know how a good draught helps a coal or wood fire to burn well and consume its fuel to ash, and we have all noticed how the flame of a gas stove becomes much hotter by letting air into the flame with an adjuster. Examples can easily be multiplied. The power for driving a car is obtained by burning fuel consisting of petrol vapour, mixed with air in an 'internal-combustion' engine. The important point to notice here is that if there were not any air the spark from the plugs could not explode the petrol vapour at all, as there would be no oxygen to join up with it. Complete combustion, we can say, calls for the provision of oxygen to join with what is being burnt or, in chemical language, combustion is the *oxidation* of fuel, with the release of energy in the shape of heat or power. It always leads to the formation of waste substances.

Active protoplasm of cells supplies itself with the energy needed by a kind of slow combustion, oxidising its fuel in the guise of sugars and starches called by the general name of *carbohydrates*. Both animals and plants are similar in the need of oxygen in order that their carbohydrates may be completely 'burned up'; this biochemical process so closely related to combustion is called *respiration*, and it likewise causes waste substances to be formed, such as carbon dioxide and water, which are useless or even injurious to the protoplasm. We may define it thus: respiration is the oxidation of 'food-fuel', resulting in a release of energy for living. (Do not confuse this particular sense of 'respiration' with its more general sense of 'breathing'.)

Every living thing needs energy in the forms of work-energy or heat-energy, or both. Even when we lie sleeping, energy is needed to keep the heart beating, the lungs moving and the body warm. Of course, a plant does not need so much energy as an animal, for it does not move about from place to place, but its life process, particularly of growth, needs energy for it to be carried on. Because plants give out oxygen as they take the carbon from the carbon dioxide around them in the sunlight to manufacture food, it is easy to forget that they, too, need oxygen for respiration and cannot live without it. A plant kept in the dark without oxygen dies. Like animals, they oxidise carbohydrates to obtain the energy they need.

One of the main differences between plants and animals is that humans and animals have to find their carbohydrates ready-made, whereas plants have the ability of manufacturing their own from the basic ingredients of atmospheric carbon dioxide and water. This calls for energy, and it is obtained from the sun, a fact which explains why this process, called 'photosynthesis', does not take place in the dark. To obtain light-energy, the green substance called *chlorophyll* is necessary to trap it. Fungi that lack chlorophyll, such as yeasts, cannot therefore work in this way, and resemble man in the respect that carbohydrates must be provided ready-made for them to oxidise.

The Carbon Cycle
(The circulation of carbon in nature)

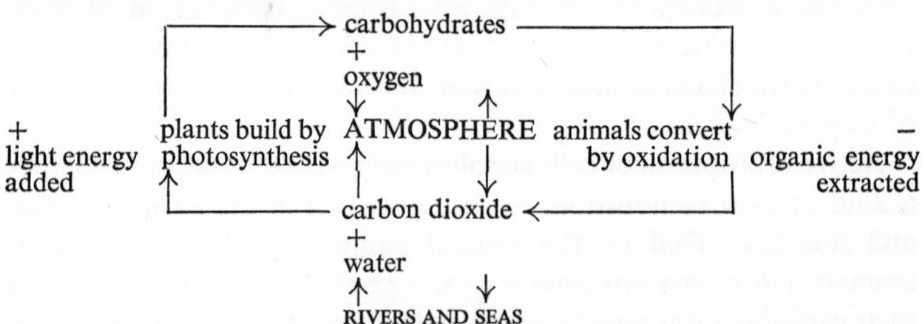

The amount of energy to keep the vital functions of the human body working is called the 'Basal Metabolic Rate', and for the average man, lying in bed, is about 1600 large calories. A 'calorie', the unit used, is the quantity of heat required to raise the temperature of 1 g of water from 15° to 16° C. A 'large calorie' or 'kilogram-calorie' equals 1000 gram-calories. To form one gram-molecule of sucrose in sugar cane or sugar beet needs about 1350 large calories of energy, which the plants trap as explained from the radiant energy of the sun.

$$12CO_2 + 11H_2O + 1349{\cdot}6 \text{ kg-cal} \longrightarrow C_{12}H_{22}O_{11} + 12O_2$$

carbon dioxide + *water* + *sun-energy* *sucrose* + *oxygen*

It has been calculated that one pound of sugar has enough organic energy in it to raise some 4½ gallons of water from normal room temperature to boiling point. All this organic energy is available to animals and man when it is oxidised back eventually to carbon dioxide and water. Wood, coal and oil fuels are carbon substances derived from dead cells of past eras, and carbohydrate 'fuels' are carbon compounds too, combined with hydrogen and oxygen. Of course, the temperature of cells is not high enough to 'burn' the carbohydrates in the general sense of the word; the carbon compounds serve as fuel because by means of the organic catalysis of enzymes they can be broken up into smaller units and fragments, and this severing of the bonds that link the atoms of the carbon compounds together releases the energy they contain. The plants that manufactured the starches and sugars trapped energy from the sun in order to link the atoms together, and now animals—and yeasts too—by using their enzymes to break the links are able to secure this energy for their own use. Ultimately, therefore, all animals are dependent on plants for their existence.

In man and animals the process of rendering food soluble and assimilable by means of enzymes, prior to the ultimate extraction of energy by oxidation, is covered by the general term of 'digestion'.

Human Digestion of Carbohydrates

If the carbohydrates are taken by animals and humans in the form of *glucose* or *fructose* they are absorbed at once and need no digestion; they can be oxidised by the respiration process without prior simplification. Other sugars, viz. cane or beet sugar (*sucrose*), malt sugar (*maltose*) and milk sugar (*lactose*), need digestion or breaking down into simpler forms, and this takes place in the small intestine, which contains the three necessary enzymes: *invertase* or *sucrase*, *maltase* and *lactase*. If the carbohydrate is starch, considerable digestive action is necessary. Cooking of some sort breaks up the natural starch grains, allowing the digestive hydrolases to have access to the starch. First an enzyme in the saliva, formerly called *ptylin*, but now *salivary amylase*, from the Greek word *amylum* = 'starch', and then another in the juice of the pancreas called *pancreatic amylase*, break down the starch into *maltose*, which is then finally split up into glucose by the maltase in the intestine already mentioned.

The blood absorbs these digested or simplified sugars from the small intestine, and conveys them into the liver, where *isomerising enzymes* convert them all into the common form of glucose. The liver finally links up the glucose units, or 'polymerises' them, to form an animal starch called *glycogen*, which it then holds in stock ready for distribution as it is needed. An average man has less than one pound of glycogen stored in his body, not enough for one day's supply of energy, but fortunately fat is the long-term storehouse of energy food; excess carbohydrate is converted to fat.

Whenever any demand for glucose is made by other parts of the body, the liver is ready to supply it from its stock of glycogen via the blood stream. In the usual way, the concentration of glucose in the blood is constant, the average in humans being 0·1%, the liver supplying it at the rate at which other tissues take it up. This level of blood sugar is maintained by *hormones*, among which is *insulin*, formed in the pancreas. The disease diabetes is an underproduction of insulin, causing too high a level of glucose in the blood, accompanied by low carbohydrate metabolism.

Metabolism is the general name given to the sum of the processes by which the units of food are utilised by the cells and organs of the body. If the process is a building-up, such as the synthesis of glycogen from units of glucose, then the term *anabolism*, or *assimilation*, is used; if it is a breaking-down, such as the severing of atomic links in the glucose molecule by oxidation, then the term employed is *catabolism*, or *dissimilation*.

All the atoms which go to make up a molecule, and there are twenty-four in a glucose molecule, are held in place by means of considerable energy. It is as if springs have been compressed to get them into position; unhook the atoms, and the energy in the springs will send them flying apart. This is what the cells do to the glucose, provided that oxygen is available, as it is, by means of the lungs of animals and humans. As the blood passes through the respiratory organs, it takes in the oxygen and passes out carbon dioxide; proceeding to the tissues of the body, it provides the cells with the oxygen they need and accepts in return the carbon dioxide they have produced. The cells can now oxidise the glucose, that is they re-arrange the atoms, 'hooking up' the oxygen with the hydrogen atoms from the glucose to form water, and more oxygen with the carbon atoms from the glucose to form carbon dioxide, both waste products of this oxidation process we call respiration. The energy locked up in carbon dioxide and in water molecules is very much less indeed than that in the glucose molecule, so that a considerable amount is released by this process. This is the energy that cells tap for carrying on life, and a detailed account of the means by which this energy is captured and stored is given in the section on the chemistry of the fermentation cycle.

Yeasts and Carbohydrates

With this basic knowledge of animal digestion and respiration: we can now turn to yeasts.

Pasteur's Effect

It was Pasteur who first observed that the presence of oxygen resulted in an increase in the respiration of yeasts but a decrease in their fermentation. In other words, when the must has free access to air, yeast growth and cell multiplication are at their maximum, but at the expense of alcohol, which is at its minimum. On the other hand, when air is excluded from the must yeast growth slows down but there are increased yields of alcohol. Pasteur used the terms *aerobic* to describe the first situation, and *anaerobic*, meaning 'without air', to describe the second. The phenomenon itself, i.e. the interference with fermentation by oxygen, is now known as 'Pasteur's Effect'.

Respiration

We have seen that plants and animals need oxygen for the chemical changes that convert carbohydrates to carbon dioxide and water for the extraction of energy, and that without it they cannot continue to live. Yeasts likewise grow at their best in a situation where oxygen in solution is freely available. They are then able to carry on cellular respiration, breaking down the carbon-to-hydrogen and the carbon-to-carbon bonds where most of the chemical energy of glucose is contained. With the release of energy for maximum yeast growth and new cell production, the carbon and hydrogen atoms are linked with the oxygen to give carbon dioxide and water (CO_2 and H_2O). Air contains some 21% of oxygen, and it is soluble in water, so that when the surface of water is in contact with air it always contains some oxygen. A litre of air contains 210 c.c. of oxygen, and a litre of water can contain 5–10 c.c. of oxygen at normal temperature, a small amount, but sufficient for the yeast's respiration process. Naturally, for the *complete* oxidation of quantities of sugar in solution, air would need to be bubbled through the must as it is in factories that produce bakery yeast on a commercial scale, otherwise cells nearest the surface take up the oxygen faster than it can diffuse into the deeper parts. The equation below assumes that ample supplies of oxygen are available. Notice the large amount of energy that the yeast is able to tap, and also that under such ideal conditions it is presumed that no alcohol is produced:

Aerobic Oxidation of Glucose: Respiration

$$C_6H_{12}O_6 + 6O_2 \longrightarrow 6H_2O + 6CO_2 + 700 \text{ kg-cal}$$

glucose *oxygen* *water* *carbon dioxide* *aerobic yield of energy*

Fermentation

It is fortunate that yeast is *facultatively anaerobic*, which means that an alternative mode of life is available for it if oxygen is not present. We may be delighted that life along these lines produces a different residual substance—alcohol, and in copious quantities, but from the point of view of the yeast, it is a less profitable way of life, for two reasons.

First, very little energy from the breakdown of sugar is able to be tapped, not 10% of that formerly available; the remaining 90% of the carbohydrate-energy, in the absence of ample atmospheric oxygen to assist in obtaining it, is left behind in the residual product. Alcohol is a powerful source of energy, and it can now be seen that this is because it retains the energy of the sugar that the yeast in anaerobic conditions is unable to reach. In other words, alcohol is an incompletely oxidised substance. Secondly, instead of the yeast continuing to develop, the point must be reached, if the fermentation continues, when some 16%+ of alcohol exists in the must. This, it must be remembered, is a waste product from the point of view of the yeast, and as it cannot be dispersed as is the carbon dioxide gas, the yeast is finally inhibited from further activity by its presence; general toxicity is the overriding factor here. In the following equation for this partial oxidation of sugar in anaerobic conditions, notice the small amount of energy now available compared with that given in the equation above:

Anaerobic Oxidation of Glucose: Fermentation

$$C_6H_{12}O_6 \rightarrow 2C_2H_5OH + 2CO_2 + 66 \text{ kg-cal}$$

glucose *alcoho* *carbon dioxide*

Yeast substance

When an air-lock is put on a vessel containing yeast with sugar in solution, and the conditions thereby changed from aerobic (ideally at any rate) to anaerobic, the yeast cells continue to multiply at the same rate as before until the oxygen previously dissolved in the liquid is exhausted, and the stimulation resulting from its presence is no longer maintained in the cells. After this, not only is the extent of their growth severely curtailed but also their rate of growth falls away. If the nitrogen content is satisfactory, then the growth of new cells about equals the number that die, so that the figure remains roughly constant during the period of fermentation.

In aerobic conditions the presence of oxygen results in very much greater yields of yeast. In the commercial manufacture of yeast for use in baking, alcohol production is almost entirely prevented by vigorous aeration, and by restricting the entry of the wort so that sugar is not present in excess, a process known as *exponential feeding*. In such ideal conditions of respiration the yield of wet yeast is 100% of the weight of molasses used as a carbohydrate source, and may even be as high as 115%. As 100 parts of molasses contain about 50 parts of invert sugar, and 100 parts of pressed yeast, because of its water content, is equivalent to 25 parts of yeast dried matter, we can take the yield of dried yeast as rather more than 50% of the invert sugar consumed.

The formula usually given for anaerobic oxidation is:

$$C_6H_{12}O_6 + 6O_2 \rightarrow 6CO_2 + 6H_2O$$

but this makes no allowance for this high production of yeast substance. Yeast dried matter contains on average in its carbohydrate and protein contents 47% carbon. It occurs in the structural cellulose and carbon skeleton of the cell, in the proteins and nucleoproteins, in the metabolic carbohydrates and glycogen, and in the yeast 'gums'. Altogether a number of the carbon atoms of the sugar are utilised in this way, and not all are given off in the form of carbon dioxide as the formula above implies. It has been suggested that a more meaningful formula in this respect would be:

$$C_6H_{12}O_6 + 2O_2 \rightarrow 2CO_2 + 2H_2O + 4(CH_2O)$$

where the four carbon atoms needed for alcohol in the fermentation formula are utilised for yeast substance (represented by CH_2O) in respiration. This would be supported by the claim that the amount of carbon dioxide given off is the same in both aerobic and anaerobic conditions.

As our study is the production of alcohol and not the manufacture of baker's yeast, we are not concerned with the pathway followed by yeast in these aerobic changes, nor with the means by which the carbon atoms are assimilated and yeast substance synthesised. There are still many problems and difficulties attached to the explanation of the mechanism of the Pasteur effect and to the understanding of the dismembering of the carbon skeleton in respiration, and the situation remains open to further research and investigation.

B. Proteins

We can now leave carbohydrates and turn to the second purpose of nutrition in living organisms, viz. to supply material for the building and repairing of cells. The carbon of sugar may provide the 'backbone', but cell protoplasm is essentially based on protein, and this is the substance for study in this section.

It was Gerardus Mulder, a Dutch chemist, who in the nineteenth century gave the name *Protein* to a group of biological substances, taking the term from the Greek word *proteios*='primary' because of their primary importance. The name chosen was a suitable one, for although a number of substances go to make up the living cell and tissue, proteins form such a large part of these that we may look on proteins as particularly characteristic of living stuff and the material from which it is made.

A number of properties distinguish them from carbohydrates and fats, but in particular their nitrogen content. Although there are many different types of proteins, all of them contain the three kinds of elements that are found in carbohydrates, carbon, hydrogen and oxygen, *plus* one never seen in starches and sugars (excluding the compounds loosely called amino sugars), viz. nitrogen. A general analysis of a protein would show:
carbon 51–55% oxygen 21–24% hydrogen 6·5–7·3% nitrogen 15–18% and often small amounts of sulphur from 0·5 to 2·5%, and less frequently traces of phosphorus.

Proteins are very numerous, and they exist in different forms. There are, for example, hundreds of proteins in human blood plasma, each with its particular job to fulfil. Every species of organism has its characteristic proteins shared by no other, and it is thought that each individual organism may likewise possess proteins repeated in no other. Sensitisation to unfamiliar proteins causes unfortunate reactions, such as nettle rash following raspberries on the menu, and hay fever from plant pollen. As each species contains several hundred different varieties of proteins, one has only to think of the thousands of animal species, each with its individual set of proteins, to realise the staggering number of different proteins in existence.

This almost endless variety of protein types is possible because proteins are constructed from what we can call 'basic building blocks'. Protein molecules are enormous in the world of cells, giants called *macromolecules*, consisting of thousands or even millions of atoms 'hooked up' together. These basic building blocks from which they are built are the *amino acids*, in themselves comparatively simple molecules containing 10–27 atoms each. Some twenty-odd amino acids have so far been discovered by hydrolysis, and if one takes 500 amino-acid units as an average for the construction of one protein molecule it is easy to see the endless combinations and permutations of patterns that are possible. Not every protein contains all twenty amino acids, but even so, the reason behind such a variety of protein types becomes clear.

Cells build up their individual requirement of proteins from amino-acid units as part of their metabolism, linking them together in the shape of strings, or branches, or coiled into balls, so that a supply of amino acids is an essential part of food supply. One of the main differences between plants and animals is that green plants are peculiarly capable of *synthesising*, or manufacturing, all their amino-acid requirements from the nitrogen in the *inorganic* forms of salts in the soil, mainly *nitrates*, together with water and carbon dioxide of the air. Man and animals, on the other hand, cannot do this, but are forced to obtain their amino acids from complex *organic* sources; herbivorous animals feed on plant life, and carnivorous animals in turn kill and eat their herbivorous relatives or one another, and then proceed to break down by digestion this protein food.

It is one of the puzzles of life that not even plants can make any use, or more correctly any *direct* use, of the enormous amount of nitrogen present in the atmosphere and forming some four-fifths of its content, in the way that oxygen is utilised for respiration. The only way in which plants can avail themselves of it is when it is 'fixed' by certain soil bacteria, such as the nitrogen-fixing species of *Clostridium* and various species of *Azotobacter*, that is, converted into chemical compounds which can be taken in by the plant roots in solution.

Man normally includes both animal and plant food in his diet, and as regards about half of the twenty-odd amino acids these *must* be included in his diet, for his body cannot synthesise them, and for their supply he is dependent on his food. Any of the others making up the twenty he is able to manufacture by the aid of enzymes, should they be needed for building up his individual kinds of protein material, and therefore it is of no importance whether they are included in his food or not. Consequently, amino acids are divided into two groups, *essential* amino acids and *non-essential* amino acids.

Most proteins usually contain all the amino acids in varying amounts, but there are some oddly deficient in one or more of the essential ones, and some proteins with large proportions of non-essential acids. Vegetable protein contains the essential acids, but on the whole a larger quantity of vegetable-protein needs to be eaten to supply the same amounts of amino acids available in smaller quantities of animal-protein. Grass contains about 10% protein, and occasionally one hears of a person claiming to live on it as food! Grasses do indeed contain all essential acids found in animal-protein, but such large quantities would need to be eaten, and the bulk of cellulose would impair digestion, for neither man nor non-ruminant animals have enzymes to digest cellulose. Yeast protein contains all the essential amino acids, although rather low in those containing sulphur, viz. cystine and methionine.

Amino acids

Essential		*Non-essential*
Arginine	Methionine	Alanine
Histidine	Phenylalanine	Aspartic acid
Leucine	Threonine	Glutamic acid
Isoleucine	Tryptophan	Glycine
Lysine	Valine	Hydroxyproline
		Proline
		Serine

Tyrosine can be made from phenylalanine
Cysteine can be made from methionine
Cystine can be made from methionine

The Chemistry of Amino Acids

The distinctive feature of amino acids is the amino group NH_2 and the acid group COOH (called a *carboxyl* group), both attached to the same carbon atom. Thus a typical amino acid formula will be:

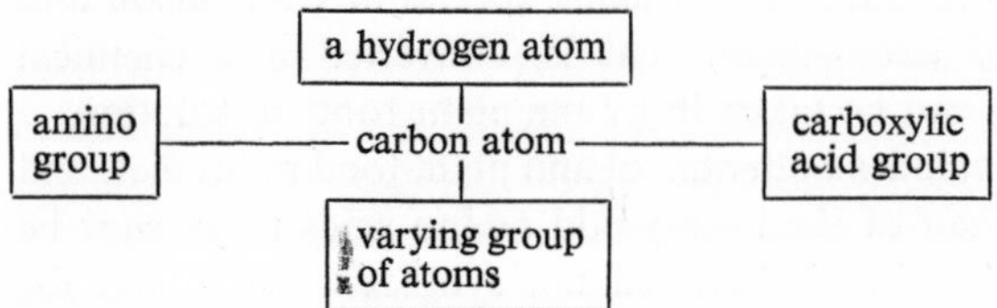

Or the same in chemical symbols, where R represents the variable parts of the acid:

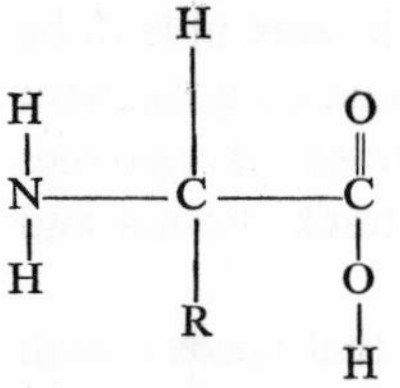

```
          H
          |
H         |       O
|         |       ||
N---------C-------C
|         |       |
H         |       O
          R       |
                  H
```

Here is the formula of *glycine*, the simplest amino acid, that can be obtained from the simplest fatty acid, acetic acid, by substituting an amino group for a hydrogen atom, although it is not always formed in this way:

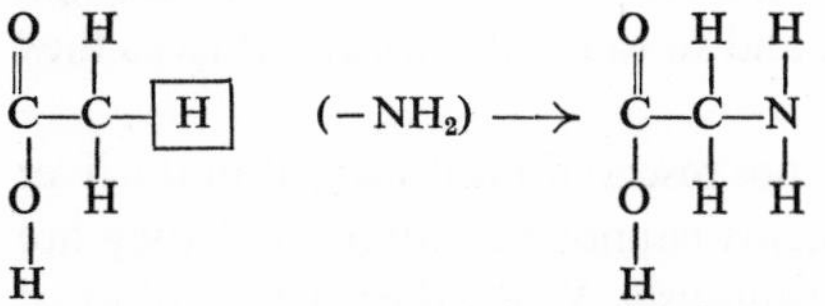

```
O  H                             O  H  H
|| |                             || |  |
C--C-[H]      (-NH2)  --->       C--C--N
|  |                             |  |  |
O  H                             O  H  H
|                                |
H                                H
```

acetic acid *amino group* *glycine*

The most important feature of amino acids is the way in which the amino group, or the amino *radical* as it is usually called, of one acid can react with the carboxyl radical of another, with condensation of water, thus forming linked chains to form ultimately protein molecules. This synthesis, with resultant dehydration as a water molecule is separated, is shown below:

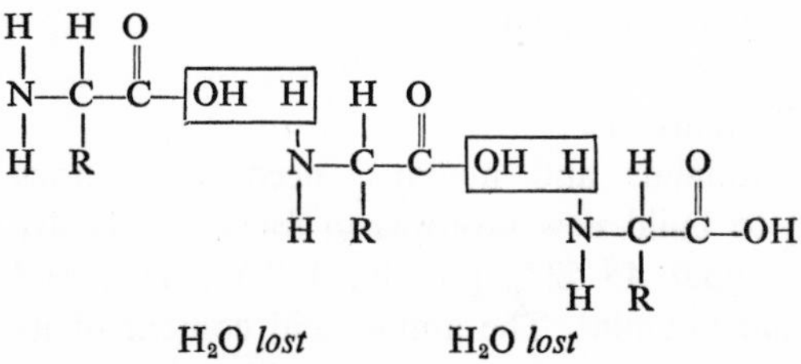

```
H  H  O
|  |  ||
N--C--C-[OH  H]  H  O
|  |         |   |  ||
H  R         N---C--C-[OH  H]  H  O
             |   |         |   |  ||
             H   R         N---C--C--OH
                           |   |
                           H   R
        H2O lost      H2O lost
```

When these two molecules of water are split off and the three amino acids linked we have a *tripeptide*. The stages in the synthesis of a protein molecule are: *amino acids*, *dipeptide*, *tripeptide*, *polypeptide*, *peptone*, *proteose*, *metaprotein*, *protein*. This head-to-tail linkage of amino acids, -CO-NH-, is known as a *peptide linkage*.

Human Digestion of Protein

Protein eaten by man and animals is broken down by digestion and the amino acids rearranged to form the type of protein required by the individual. Even if the large protein molecule were able to be absorbed *in toto*, it would be unsuitable, for it would be a different variety and not man-protein. Therefore a complex of enzymes called *proteolytic* enzymes exists in the body to attack protein and break it down into its basic units.

First is *pepsin* in the stomach juices, an enzyme that needs a high acid condition for activity. Next follows the *pancreatin* group of enzymes in the pancreatic juice, and finally, a group of intestinal enzymes called *erepsin*. Each of these enzymes has the task of attacking the long-linked protein chain of molecules at specific points, so that it is broken down into shorter and shorter links, until finally the unit blocks of amino acids are unlinked, absorbed through the small intestine into the blood stream, and so to the tissues and cells as they are needed.

Unlike carbohydrates, which are reserved until wanted in the liver or as fat about the body, proteins cannot be stored, and they are broken down and built up continuously. When there is a surplus of protein in the diet the amino acids are *deaminated*, that is they lose their amino groups. The nitrogen of this group is removed as ammonia, a highly toxic substance for animals and humans, and therefore it is converted to a non-toxic form called *urea*, and excreted in the urine. The fatty *keto acids* left after deamination are either stored with the carbohydrates and fats or oxidised to carbon dioxide and water to provide energy. In this way animals and humans, who cannot make use of inorganic compounds of nitrogen as do plants, utilise safely their surplus protein.

Yeast Assimilation of Proteins

Yeasts, like plants, animals and humans, need nitrogenous materials out of which to build new protoplasm for cells. As dry material, they contain about 45–52% protein, 1–2·6% fats, and various mineral salts and vitamins. The amino-acid content of its protein compares well with muscle protein although low in methionine, and indeed yeast protein is placed between that of animals and vegetables for biological value. The protein metabolism of yeast is very accommodating, with few of the restrictions that we have seen to apply to plants and animals, and we can consider its protein nutrition supply from three sources:

1. *Protein*

Yeasts contain a full complement of proteolytic enzymes capable of breaking down protein substances into units of the size required, but it has been found, especially by research carried out on panary fermentation, that yeasts do not secrete *exocellular proteases*, that is, they do not pass proteolytic enzymes outside into the must or wort, so that any form of protein material available must be broken down *inside* the yeast cell. Protein molecules, however, are far too large to diffuse through the semi-permeable membrane of the cell wall, and even the polypeptides are unable to enter (see notes on diffusion). The largest molecules capable of passing inside are the simpler peptides, and therefore where protein material is supplied in the natural form of cereal, grape, fruit or other ingredient with a protein content, the yeast is dependent for its nutrition on the proteinase enzymes of the ingredients themselves to break down the long chain into peptide units small enough to diffuse into the cell. There is one exception to this restriction. During a long fermentation *autolysis* may take place, a process in which the cells that have completed their life-cycle are broken down by their own intracellular enzymes, for proteases are quite capable of attacking the cell material and other enzymes when out of control, and thereby escaping beyond the confines of the cell. In this way, *proteinase*, *polypeptidase* and *dipeptidase* (yeast *erepsin*) are released to attack proteins externally.

2. *Amino Acids*

Tripeptides and dipeptides pass easily into the cell, where they are hydrolysed by *peptidases* into their constituent amino acids, so that we can go on to discuss the treatment of the latter by the yeast without further consideration of these larger units.

The German chemist, F. Ehrlich, was the first in 1907 to elucidate this problem. He declared that they were hydrolytically *deaminated* (the amino group removed), as with animals that have surplus supplies of protein, but, quite unlike animals, the resultant ammonia, an excellent nitrogenous source, was then assimilated into the cell and the residue *decarboxylated* (the carbon dioxide removed) into a primary alcohol. The enzymes taking part in these changes are called *deaminases* and *carboxylases* respectively, both in abundant supply in yeasts. No ammonia remains in the medium; it is assimilated as soon as it is produced.

ASSIMILATION OF AN AMINO ACID BY YEAST

A. Deamination

$R{\cdot}CH(NH_2)COOH + H_2O \rightarrow R{\cdot}CH(OH){\cdot}COOH + NH_3$

the amino group — *water for hydrolysis* — *ammonia assimilated by yeast*

B. Decarboxylation

$R{\cdot}CH(OH){\cdot}COOH \rightarrow R{\cdot}CH_2OH + CO_2$

residue after deamination — *higher alcohol* — *carbon dioxide removed*

In 1949 R. S. W. Thorne, an English specialist in yeast, working under the Institute of Brewing Research Scheme, produced evidence that suggests that yeasts can absorb their amino acids as *intact units* into their cell-substance protein, and only where certain amino acids are missing or deficient do yeasts fall back on their ability to break down by Ehrlich's mechanism any available amino acids in order to obtain the ammonia by means of which they can synthesise the deficiencies. He has been able to grade many amino acids into good, average and poor nutrients, dependent on the relative ease of deamination, and has found that a mixture of them is preferable to one alone, a solution of seven plus ammonium phosphate being not far behind the nutrition obtained from normal brewer's malt wort.

Probably in practice, both methods of obtaining nitrogen are utilised by the yeast, and Thorne himself suggested that 50% of its nitrogen intake is from intact amino-acid units, and 40% from the ammonia formed by their deamination. The final 10%, he thought, came from an alternation of oxidation and reduction between pairs of amino acids, one being a donator of hydrogen, and the other an acceptor. This process, known as the 'Strickland reaction', releases ammonia as follows:

THE STRICKLAND REACTION

A. Hydrogen donor

$R{\cdot}CH(NH_2)COOH + 2H_2O \rightarrow R{\cdot}COOH + NH_3 + CO_2 + 4H$

ammonia released

B. hydrogen acceptor

$2R{\cdot}CH(NH_2)COOH + 4H \rightarrow 2R{\cdot}CH_2COOH + 2NH_3$

hydrogen donated by one amino acid to another

SUMMARY OF AMINO ACID ASSIMILATION

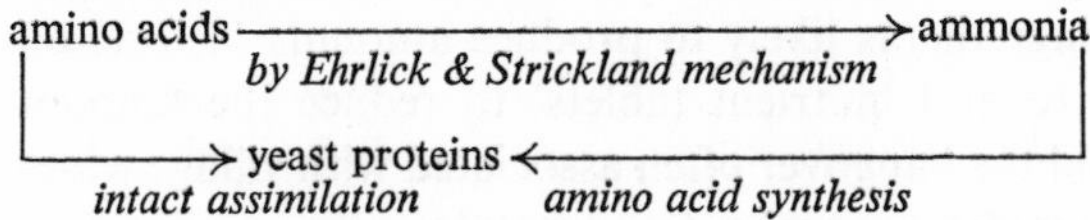

The higher alcohols, formed as Ehrlich has described, are waste products from the yeast's point of view, but in wine they are of immense importance, for though appearing only in traces, they intermarry among the acids present to form the esters that bring bouquet and flavour to wine during its maturation.

3. *Ammonium Salts*

Like green plants, but quite unlike animals, yeasts can utilise ammonium salts as a source of nitrogen, though only certain genera can make use of nitrates, and *S. cerevisiae* is not one of these. From such salts, yeasts build up their amino acids, and with these units go on to construct yeast-protein.

It may be true, as Thorne has observed, that mixtures of amino acids are better nitrogenous nutrients than ammonia, and also that where the latter is the sole source of nutrient more carbohydrate will be diverted to provide the carbon skeleton for the manufacture of amino-acid units than where these are assimilated intact, but nevertheless ammonium salts in the shape of 'nutrient tablets' are the normal way in which winemakers supplement protein material. Cereals, such as barley for beer, and pure grape juice for wine, bring ample constituent supplies of protein for yeast, but where juices are diluted, and this is usually the case with home-made wines, it may be necessary to supplement them with ammonium phosphate or ammonium sulphate, the basis of 'nutrient salts' supplied commercially. Where protein is deficient, as with honey for mead making, or completely lacking, as with flowers for 'flower wines', then such salts are essential additives to the must. Alternatively, a cereal or raisins can be included in the recipe to remedy this lack of protein in the ingredients.

There is a further situation when it is advisable to add an easily accessible source of nitrogen. The amino acids *leucine* and *isoleucine*, which are both contained in potatoes and cereals, are broken down by Ehrlich's mechanism, leaving a residue of *isoamyl* and *active amyl*, the main ingredients of *fusel oil*. Harmless in traces and beneficial to the flavour of some wines, fusel oil in excess becomes toxic to the drinker. Ehrlich noticed that the addition of ammonia depressed the production of fusel oil, suggesting that the yeasts were assimilating their nitrogen in the form of ammonia. Winemakers, therefore, who make wines from ingredients likely to produce amounts of fusel oil would be advised to add 'nutrient tablets' to reduce the amount formed and so avoid the hangover often associated with village wines made from such ingredients as wheat and potato.

Whatever the source of nitrogen, whether protein, amino acid or ammonium salts, yeasts are capable of synthesising any deficiency in amino acids; there are no essential amino acids as with humans, and growth is not dependent on a supply of certain of these being present.

C. Vitamins and Mineral Salts

The main needs of yeast are for the elements carbon, hydrogen, oxygen and nitrogen. Water can supply the hydrogen and oxygen, so that the major components for the growth of yeast cells are carbon and nitrogen, and the assimilation of sugar and protein can provide these along the lines discussed. Although these four elements constitute anything up to 95% of the content of dried yeast cells, there are still certain other substances essential for life in yeast, just as there are in man and the higher animals, although they need be present only in minute amounts. Deficiencies result not only in a reduced rate of growth but also in a weakening of fermenting power. Certain minerals, in the form of ions from their soluble salts, must be present for the activating and functioning of some enzyme systems, and they are then known as *cofactors*. Certain vitamins are essential constituents of some coenzymes: nicotinamide enters into formation of the 'cozymase' molecule, phosphorylated thiamine is the coenzyme of carboxylase, pantothenic acid is the precursor of coenzyme A. Thus minerals and vitamins are vital to the processes of metabolism in yeast as in humans.

Vitamins

It was Casimir Funk, in 1911, who cured polyneuritis in chickens by feeding them with a substance that he extracted from rice bran with dilute acid. He decided that it must be one of the organic compounds based on ammonia called *amines*, and as it was evidently necessary for life, he added the word *vita*='life', and so provided the word *vitamine*, later varied to *vitamin* to avoid confusion with *amine*. As vitamins were discovered, they were distinguished by letters, A, B and so on, but now that they have been chemically determined they have assumed individual names, and the alphabetical classification is not really needed. They fall into two main groups:

Fat-soluble: containing only carbon, hydrogen and oxygen; A, D, E and K types.

Water-soluble: most also contain nitrogen; B and C types.

Pasteur in his researches showed that yeast could be grown on a medium of sugar, ammonium tartrate and yeast ash. These provided respectively the carbohydrates, the nitrogen and the mineral salts needed for growth. But later at the turn of the century, E. Wildiers, in 1901, failed to grow a 'very small seedling' of yeast on this medium, whereas a larger one succeeded. He concluded that the latter was large enough to bring something with it from the original culture, whereas the smaller one did not, and therefore some additional unknown substance was essential for growth. He called this *bios*, and no further attention was given to it. Then, in 1919, an American R. J. Williams linked bios with the vitamins, and we now know that these belong to the B-group of vitamins, most of them identical with the vitamins of the higher animals. The most important for yeast are:

biotin	pantothenic acid
inositol	pyridoxine
nicotinamide	thiamine (aneurine)
riboflavin	

A yeast's need for certain of these vitamins is known as its *bios factor* or *growth factor* requirement, and the result of their stimulation on the cells is the *bios effect*. Yeasts vary very much in this respect. Presumably in their original wild state they were able to synthesise all their requirements, but generations of culture in nutrilite-rich medium has weakened their power and made them dependent to some extent on external supply. For this reason, living yeasts, far from providing vitamin B, may actually deprive the consumer of what is present in his intestines by absorbing it themselves if they are deficient in this respect.

All genera and species of yeasts can synthesise *riboflavin*, particularly the lactose-fermenting yeasts, although for its commercial production *Eromothecium ashbyii* is utilised. Some, such as the 'food-yeast' *Candida utilis*, can synthesise all the other vitamins mentioned above. On the other hand, others, such as *Kloeckera brevis*, can synthesise none of them, and need to have all supplied in the medium. Most yeasts are somewhere between these two extremes, and as even strains of the same variety may differ in their vitamin requirements due to conditions of growth, it is not possible to say accurately what these are in general for *Saccharomyces cerevisiae*. Some indication, however, may be given by the result of an experiment conducted by Burkholder, McVeigh and Moyer in 1944 with 163 strains of yeasts that need a wide range of vitamins. Of the seven named above as important for yeast, the deficiencies in order were: biotin 78, thiamine 33, pantothenic acid 30, inositol 15, pyridoxine 13, nicotinamide 13, riboflavin 0.

The value of yeast as a supplementary source of B-group vitamins for humans is obvious, and readers will know of its medicinal use in this respect. Comments on the protein food-value of yeast have been given in an earlier page.

Minerals

Mineral salts are also needed by yeast cells, but in the merest trace amounts; in larger quantities they are capable of having a reverse effect and inhibiting growth. The most important are:

Phosphates

Harden, the English chemist, showed early in the century how important these are for yeast activity. They play a vital role in the mechanism of carbohydrate metabolism, making energy available to the yeast, and also enter into the manufacture of nucleoplasm.

Sulphates

These are needed for protein synthesis, for certain of the amino acids contain sulphur in their make-up, notably cysteine, cystine and methionine. The sulphur is probably assimilated by reduction of the SO_4 via hydrogen sulphide H_2S.

Potassium

This chemical is connected with the process of respiration, and its ions stimulate the enzyme *enolase*.

Magnesium

This enters into the manufacture of nucleoplasm, and its ions activate enzymes concerned with phosphate transference.

Trace amounts of other ions, such as iron, copper, zinc and manganese, are all capable of stimulating growth. Calcium ions also activate enzymes, although unessential.

A Nutrient Formula

Grape juice is the ideal medium for yeast fermentation, but most fruits are quite satisfactory in this way. If, however, the must is felt to be deficient in nutrition a little malt extract with a trace of Marmite assists, or use can be made of nutrient salts. For those who wish to make up their own, the following recipe adapted from that of the British Bee Keepers' Association is basically sound:

Ammonium sulphate	4 g
Magnesium sulphate	½ g
Potassium phosphate	2 g

This is sufficient for 2 gallons of wine, and supplies nitrogen together with four essential elements.

Section 2 *Sugars*

Chapter 5 Sources of Supply

Although the Moors introduced the sugar cane to Europe in the eighth century, there was little effect on the daily diet, for the process of extracting and crystallising the juice was slow in being understood, and honey remained the universal source of sweetness from the earliest times up to the end of the Middle Ages.

The request by Henry III to the Mayor of Winchester to send him three pounds of sugar, 'if so much could be had', is often quoted as an illustration of the scarcity of sugar in those early days of its introduction. Even when it began to arrive in this country in some quantity, from Venice at the beginning of the fourteenth century, it was still very much a luxury, and if by the end of that century it had become a familiar article of food it was still far too expensive for the general public to purchase for regular consumption.

Advances in means of transport, extension of trade relations and the development of the sugar industry in the West Indies brought a sharp drop in prices and a remarkable increase in sugar consumption, so that from an annual consumption-rate of 4 lb a head in 1700, 18 lb in 1800, the figure has steadily risen to over 100 lb at the present day.

Until the eighteenth century the sugar cane was the normal source of sugar. In appearance it looks like a giant bamboo, but it is really a species of perennial giant grass called *Saccharum officinarum*. It needs a tropical or sub-tropical climate for growth, such as Europe cannot supply. We have seen in the first section that all plants contain a certain amount of sugar, formed by the chlorophyll of their leaves, and as early as 1747 a German scientist, Andreas Maggraf, had called attention to the large quantity contained by the white beet named *Beta maritima*, for it flourishes wild on the shores of the Mediterranean. It was not until the Napoleonic wars, however, when supplies of cane sugar were interrupted, that the French started cultivating the sugar beet in earnest as a local substitute. Britain showed little interest until the U-boat blockade made itself felt in the First World War, and then the value of home-produced sugar became obvious. As a result, East Anglia now grows large quantities of this crop, and the European sugar beet now accounts for some 35% of the world's supply of sugar, as against 65% from the tropical sugar cane, and the general conclusion is that sugar can be produced almost anywhere in the world.

Other sources of supply are slight indeed when compared with these two. In Canada the sugar maple, *Acer saccharum*, is tapped for maple syrup, a delicious accompaniment to waffles, but as only about 2 lb of sugar is obtained per tree per year, this form of sweetness is something of a luxury. Some varieties of palm trees, such as the Coquito Palm, are tapped in a similar way to produce a sugar called *akrah* or *dulloah*, but this practice is confined to localities in the Orient. Experiments have been made recently with another grass that stores sugar, called *Sorghum*, especially in Russia and the United States, but difficulties presented by large constituent amounts of starch and gums still remain to be overcome if refined sugar is to be obtained from it. None of these sources presents sugar for the world's markets.

From a chemical point of view, all refined commercial sugar is identical, given the name of *sucrose* and the formula $C_{12}H_{22}O_{11}$, but it is quite natural that we should continue to use the names cane sugar and beet sugar according as the sugar cane or the sugar beet is the origin of the sucrose. Winemakers, who need to *chaptalise*, or supplement with sugar, their musts, usually express a preference for one or the other, despite the assurance of their scientific cousins that there is no difference after refining. Perhaps the leaning towards cane sugar has a psychological cause, based on feelings such as the sugar cane being the original source, with sugar beet being utilised as a substitute and so on. It is a matter for the individual winemaker to decide for himself, although it is worth noting that champagne manufacturers refuse to use beet sugar for their secondary fermentation in bottle 'because of its earthy flavour'.

The Manufacture of Sugar

1. Cane Sugar

When the 'tassel', a plume of flowers, appears at the top of the stalks the sugar canes are cut by hand with a *machete*. The sugar juice is stored by the plant in its solid stem of cellular tissue, and it is an interesting point that if the canes are cut before they are fully ripe large quantities of *invert sugar* are present, an undesirable state of affairs for the producers. At the sugar factory five stages are necessary in the manufacture of sugar.

Juice Extraction

The cane passes through sets of revolving knives, and then between crushing and squeezing rollers that extract the juice under tremendous pressure. It is dark brown in colour, with an acid content of pH 4·5–5·5, and contains amounts of pectin, gum and suspended matter. The sucrose content of the juice is about 12–18%.

Purification of the Juice

The acidity of the juice must be neutralised, or the sucrose would break up into the two simpler sugars, *glucose* and *fructose*, when heated. Further, as some of the non-sugar impurities are dissolved, they cannot be filtered but must be precipitated by chemical action, i.e. thrown out of the liquid as a sediment or scum. The next stage, therefore, is to add milk of lime (slaked lime in water) and heat the juice to boiling point. The clear juice can later be drawn off from the settlings. If a more thorough purification is needed sulphur dioxide is pumped through the liquid, and the lime is precipitated as calcium sulphite.

$$Ca(OH)_2 + SO_2 \longrightarrow CaSO_3 + H_2O$$

calcium hydroxide (slaked lime) + *sulphur dioxide* → *calcium sulphite* + *water*

Concentration of Juice to Syrup

Now the juice can be thickened by evaporation of water, and this is done by boiling in closed pans under vacuum. By this means the boiling point is lowered and the harmful effect of high temperatures on the sucrose is avoided.

Crystallisation of the Syrup

This is really another stage of evaporation in closed vacuum pans. The thick dark syrup is boiled until minute crystals appear, and then these are allowed to build up until the pan is full. The result is a mixture of some 65% crystals and 35% syrup, the latter known as the 'mother-liquor'. The whole mixture is termed *massecuite*, a French word meaning 'cooked mass'.

Separation of the Crystals

Finally, the crystals are separated from the mother-liquor by rotation in a centrifuge, so that the syrup is thrown off through perforations and the crystals remain. As the mother-liquor contains a high percentage of sucrose, it has a second and a third boiling and separation treatment before it is discarded as exhausted. It is termed *molasses*, and is the source of rum when fermented.

The sugar resulting from these processes is termed 'raw sugar', containing about 95% sucrose, and it is sacked ready for transport to the sugar refiners.

The Refining of Raw Cane Sugar

On arrival in the country where the sugar is to be marketed, the raw cane sugar is taken to the refineries. There it undergoes four stages of refining that result in a product that contains 99·95% of sucrose as compared with some 95–96% in the raw sugar when it arrives.

WASHING THE RAW SUGAR. When the raw crystals arrive at the factory they are coated with adhering molasses, and to remove this they are mixed with a hot syrup that softens the film and then spun in centrifugal machines. This removes the syrup, and hot water is sprayed on to the crystals to wash off any molasses that still adhere, without appreciably dissolving the crystals themselves.

MELTING THE RAW SUGAR. Next this washed sugar is dissolved in hot water, and the 'washed raw sugar liquor' is conveyed to 'melting tanks', where it is stirred and strained.

CHARRING THE LIQUOR. The third stage is a slow percolation through tall cylindrical vessels filled with granulated bone charcoal or 'char'. Bone charcoal, made by heating bone to redness, is more efficacious than wood charcoal because of its fine subdivisions, although a wood charcoal treated with mineral salts and called 'activated', is said to be more effective still. This char not only removes organic impurities but also any colour that remains, so that the liquor that emerges is clear and colourless.

CRYSTALLISATION OF THE LIQUOR. Finally, in vacuum pans such as those used for producing raw sugar, the sugar is crystallised out, the vacuum allowing boiling point to be reduced from perhaps 250° F in open pans to 150° F so that the sugar is not discoloured by intense heat. The centrifuge separates the crystals from the mother-liquor, and after drying in the granulator the sugar is ready for grading and packing.

2. Beet Sugar

The same five stages of cane-sugar manufacture are to be distinguished, although naturally there are certain differences of process, especially in the first two stages.

Juice Extraction

The beet, containing on average 15% sucrose, are first fed into a slicing machine which cuts them into V-shaped *cossettes*, about 4–8 inches long, and $\frac{1}{16}$ inch thick. Then they are conveyed to the 'diffusion battery', which replaced rollers for pressing out juice in the middle of last century. This consists of a series of tanks through which hot water passes over the cossettes, moving from one tank to another and becoming progressively sweeter as the sugar diffuses from the beet slices into the water; the colloidal constituents, proteins and pectins, diffuse far more slowly, and so remain behind.

Purification of the Juice

The 'raw juice', as the sweet liquid is now called, containing 12–14% sugar and some 4% impurities, needs purifying. Milk of lime is added, as with sugar cane, but in this case carbon dioxide is pumped through the liquid, a process called *carbonation*. The lime is precipated as chalk, and impurities removed with it, the liquid being filtered off:

$$Ca(OH)_2 + CO_2 \longrightarrow CaCO_3 + H_2O$$

slaked lime + *carbon dioxide* → *chalk* + *water*

Final Stages

The remaining stages of concentration, crystallisation, separation of crystals and refinement of lower-grade types are all very similar to the processes described above for cane sugar, although there may be some variation in detail.

Chapter 6 Types of Sugar

A. Disaccharides or 'Two-unit' Sugars

1. Sucrose

Sucrose is marketed in many different guises, and those of interest are taken in turn.

Granulated Sugar

This is the normal white sugar that appears everyday on the household table. Its origin may be sugar cane or sugar beet, or a mixture of the two, for after refining to the extent of 99·95% purity the end products are chemically indistinguishable. This is the sugar normally used in winemaking for supplementing the 'natural' sugar of the ingredients, and supporting its choice is not only its price but also the fact that it brings no characteristic flavour or colour, and is easily converted into alcohol by the enzymes contained in the yeast cells of *Saccharomyces cerevisiae*, the yeast employed for this fermentation.

Caster Sugar

Because of its finer grain, this is suitable for use in sprinklers or casters that 'cast' the sugar over food. Like 'preserving sugar', a sugar useful for jams because of its 'porosity', it brings no advantage to the maker of wine in purity or content.

Brown Sugar

This is a product of the refineries. In the first stage of refining the raw sugar is mixed with a hot syrup to soften and remove the film of molasses on the crystals, and this 'wash syrup' grows darker as the process continues. It contains fructose and glucose, as well as sucrose, and later the sugar is crystallised out from it in the normal way. This sugar is not washed, but allowed to retain its molasses according to the shade of colour required, and sold as 'brown sugar'. To wines it brings a treacly flavour and a darkening of colour, and it is not used therefore unless a special effect is desired.

Demarara Sugar

On the other hand, this is a raw cane sugar, a fact worth noticing. Beet juice is unsuitable for the production of this, because of the objectionable flavour of the product. The name used to indicate its origin from country around the river of that name in British Guiana, but this is not necessarily so today. The cane juice is carefully purified by the use of milk of lime and sulphur dioxide, and during its second stage of evaporation to syrup, phosphoric acid (H_3PO_4) is added to lighten the dark liquid to a yellow colour. Occasionally, a dye called 'golden bloom' may be added in the vacuum pan, and the so-called demarara be merely white crystals dyed yellow. The film of syrup is allowed to remain on the crystals, and these are larger than those of granulated sugar. Again, a distinct flavour of treacle or caramel can be detected in wines made from this sugar, although not so offensive as with brown sugar, and allowance must be made for this secondary flavour.

Loaf Sugar

Originally this was the 'sugar loaf' in the shape of a cone, weighing from 3 to 7 lb. When cut up into cubes the name loaf was still applied, and it continued to be used when the cubes were manufactured from slabs. Its sparkle and convenient shape have advantages for the table but not for the winemaker.

Syrups

Molasses or 'blackstrap' is the exhausted mother-liquor, a thick uncrystallisable dark syrup, remaining as a by-product in the fourth stage of sugar manufacture described above. It contains sucrose, invert sugar, organic and inorganic impurities. Rum is distilled from a fermentation of it, and industrial alcohol is another product, but its overpowering flavour renders it quite unfit for wine production.

Treacle is a clarified form of molasses, made by filtering a diluted solution of it and then reconcentrating, and though useful for stout brewing, it has no place in wine.

Golden syrup is quite distinct from both these syrups, for it is not a by-product at all, but a specially prepared mixture of refined sucrose and invert sugar with certain non-sugars to produce a characteristic smoky taste, pleasant on bread and puddings, but hardly necessary in wine.

Invert Sugar

This is not a disaccharide but a mixture of two simple sugars.

Before the glycolytic, or fermenting, enzymes of yeast can ferment sucrose it is necessary for this to be degraded into its component simple sugars. When sucrose is used in a must (sometimes the sole source of sugar but usually supplementing the natural sugars of the ingredients) a yeast enzyme called *invertase* splits this disaccharide into the unit-sugars that constitute it, glucose and fructose, in equal quantities. The glycolytic enzymes, sometimes referred to generally as the *zymase complex*, can then proceed to convert these monosaccharides into a alcohol by the fermentation process.

$$\underset{sucrose}{C_{12}H_{22}O_{11}} + \underset{water}{H_2O} \xrightarrow[invertase]{} \underset{glucose}{C_6H_{12}O_6} + \underset{fructose}{C_6H_{12}O_6}$$

Chemists refer to this preparatory process as *hydrolysis*, as a molecule of water is absorbed in the reaction, but a more common term is *inversion*, and when simplified in this way the sucrose is said to be *inverted.* Invert sugar is therefore a mixture of equal amounts of glucose and fructose obtained by the hydrolysis of sucrose.

On a commercial scale, it is manufactured by means of heating sucrose with acid, followed by neutralisation of the acid and then filtration. This method is known as 'acid hydrolysis', and as there is acid in our must, some of the sucrose there is likely to be inverted by acid hydrolysis in addition to that which is being changed by enzymic hydrolysis, for sucrose hydrolyses much more readily than other disaccharides, such as maltose, and there is even a tendency for sucrose in an aqueous solution to hydrolyse slowly on its own account. The general result is that one can depend on all the sucrose added to a must being inverted and ready for conversion to alcohol long before the fermentation ceases its activity.

The apparently odd use of the name 'inversion' for this process refers to the effect of a solution of sugar on *polarised light*, discussed a little later in this section. Cane sugar rotates such light to the right, whereas invert sugar reflects it to the left, so that the hydrolysis of cane sugar 'inverts' the effect of its solution on polarised light. It is not the only way in which sucrose differs from its individual constituents, glucose and fructose.

In leaving sucrose, it is worth noting that when it is heated above its melting point, 160° C, it breaks up into various brown-coloured amorphous substances called *caramel.* Its use in this form as a colouring and flavouring additive to wines and spirits is well known.

2. Maltose

This is malt sugar, formed by the germination of grain, particularly barley. It is less sweet than sucrose, and breaks down into simpler sugars less readily because it has no fructose constituent. In brewing parlance, the term 'malt' is used for the malted *grains* of cereals, whereas the public normally apply the word to the brown glutinous extract from malt grain, obtained by boiling the latter in water and evaporating the solution, viz. malt extract. It is possible to purify and crystallise it, when fine white needles of malt sugar are formed. The powerful 'beery' flavour renders malt extract unsuitable as a substitute for sucrose in winemaking, although it is sometimes fermented out as the main ingredient to produce 'malt wine'. Added in very small amounts, its nitrogenous content can give encouragement to a slow fermentation.

For reasons given under 'Fermentable Sugars', yeast does not normally break down starch into a fermentable form. Fortunately, the barley ear itself contains appropriate enzymes, named by the two French chemists who discovered them in 1833 *diastase*. During the malting and mashing processes of brewing, the diastase is activated and the starch is broken down into maltose. We now know that the diastase complex contains at least two enzymes, *α-amylase*, and *β-amylase*, each of which the chemist can separate. The former breaks down starch into *dextrins*, gummy substances, used among other things for the adhesive on postage stamps, and the latter breaks down starch and some of the dextrins into maltose. The production of maltose by enzymic hydrolysis is discussed in detail in the section on alcohol.

Maltose is a disaccharide, with the same formula as sucrose, and yeast cannot ferment it until it is broken down into a simpler form. This it does by means of *maltase*, which performs a similar function for maltose as invertase does for sucrose. The result, however, is different, for the maltose molecule is split into twin molecules of glucose, and not glucose and fructose as before:

$C_{12}H_{22}O_{11}$	+ H_2O	$\longrightarrow$ $C_6H_{12}O_6$	+ $C_6H_{12}O_6$
Maltose (1 molecule)	Water	Glucose (1 molecule)	Glucose (1 molecule)

As with sucrose again, an alternative method of breaking down starch and maltose is by acid hydrolysis, and this is the method used for the commercial production of glucose. There is an important difference between the enzymic and the acid processes, and this is that if the acid conversion of starch is taken to completion it is entirely hydrolysed to glucose. In the case of breakdown by enzymes, certain dextrins are formed together with the maltose, and as these are unaffected by maltase, they remain unfermented. Incidentally in beer they are an important factor, for apart from flavour and character, they are valuable as foam stabilisers. The following table will make the difference clear:

Acid Hydrolysis

Starch $\xrightarrow{\textit{partial hydrolysis}}$ Maltose Dextrins $\xrightarrow{\textit{complete hydrolysis}}$ Glucose

Enzymic Hydrolysis

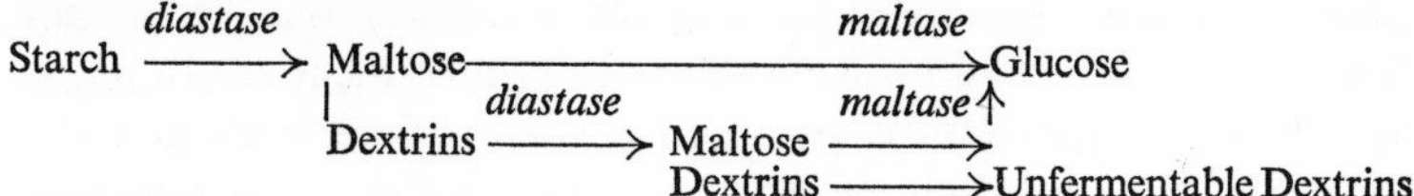

3. Lactose

Milk sugar is given this name, and it may be obtained by the evaporation of whey from which the fat has been removed. It forms small white, hard and gritty crystals that dissolve with some difficulty in water. Cow's milk contains 4·5% and human milk 7% of lactose. Like sucrose and maltose, it is a disaccharide, capable of being split into simpler forms, but unlike them, it is never found in the vegetable kingdom, and, further, its constituent sugars are equal parts of glucose and *galactose*, an *isomer* of glucose and resembling it in most respects. As with the two other disaccharides, lactose can be degraded either by acid or enzymatic hydrolysis, and the two resultant sugars can then be fermented by yeast to alcohol, together with the formation of some lactic acid.

Certain species of *Saccharomyces*, such as *S. lactis* and *S. fragilis*, are capable of breaking down lactose, but *S. cerevisiae* does not contain the necessary *lactase*, and therefore leaves it unfermented. This makes it useful for sweetening wines and beers, such as 'milk stout', where no further fermentation is desired. Otherwise it has no value for the winemaker, and indeed its sweetening power is only about one-third that of sucrose. In the East, however, it is the basis of fermented milk drinks, such as koumiss from asses' milk.

Comparative Table of Sugars

Name	*Origin*	*Component sugars*	*Hydrolytic enzyme*	*Sweetness factor*
Sucrose	Cane and beet	Glucose and fructose	Invertase	10
Maltose	Grain	Glucose	Maltase	3·5
Lactose	Milk	Glucose and galactose	Lactase	3·0

All three sugars are alike in being dextro-rotatory, and in yielding simple sugars on hydrolysis that are fermentable by *S. cerevisiae*. This yeast contains maltase and invertase, but not lactase. Maltose and lactose are 'reducing' sugars, but not sucrose.

B. Monosaccharides or 'Single-unit' Sugars

1. Glucose

So far we have been dealing with disaccharides or two-unit sugars.

Now we come to consider the simple sugars which, as can be seen in the table, form the constituent parts of these. Glucose, a most important sugar, enters into the composition of them all. In nature it occurs together with *fructose* in grapes and various fruits as well as in honey, and it is glucose that is found in the blood stream of humans and animals.

Commercially, glucose is prepared by the acid hydrolysis of starch, for this basic ingredient is not only much cheaper than any sugar, but when completely hydrolysed in this way provides glucose entirely. In America, green maize starch, in Europe, maize, rice and potato starch, are heated in closed boilers under pressure with dilute hydrochloric or sulphuric acid. After an hour and a half this liquid is neutralised with ample soda ash, and the salt formed from the acid is filtered off. The glucose solution, after partial concentration, is passed through carbon for decolorisation, and finally, following further concentration in vacuum pans, crystallisation takes place somewhat in the manner of sucrose manufacture.

The result is a white powder, composed of small six-sided crystals, with about 75% the sweetness of sucrose, an extremely pure sugar, although sometimes the glucose is left in a moist, crystalline mass without recrystallisation. It dissolves freely in both water and alcohol, and its solution rotates the plane of polarised light to the right.

This latter feature accounts for its alternative name of 'dextrose', from the Greek *dextro* = 'to the right'. In chemistry it is more exactly written '*d*-glucose' to distinguish it from a number of very similar sugars called *isomers*, and this initial letter '*d*' refers to the spatial arrangement of the atoms comprising it. In commercial practice the term 'glucose' is very loosely used, for it can refer to substances where *d*-glucose is only a proportion of the composition. The glucose used in confectionery, for example, is made from starch that has purposely been only partially hydrolysed, and consists of glucose, maltose and dextrin. Liquid glucose, known to the trade as 'corn syrup', has a little sucrose added for sweetening purposes, so that its composition may be something like: glucose 10%, maltose 30%, dextrin 50%, sucrose 10%. All are good energy foods, and only dextrin has no sweetness, but nevertheless this 'glucose' is very different from the pure *d*-glucose or dextrose just mentioned. Its use in confectionery retards the crystallisation tendency of sucrose, and brings a smooth texture to creams and elasticity to caramels. Jam manufacturers likewise find it a suitable form of sweetening.

Used in winemaking, either by itself or forming part of invert sugar, it short-circuits the need for the yeast to invert the sucrose that would normally be added to the must, and the fermentation starts without need for this 'predigestion' of the sugar. Its expense must be balanced against this, and individual winemakers should decide for themselves whether its use brings any detectable improvement in quality, which is, after all, the touchstone for its employment.

2. Fructose

This is sometimes called 'fruit sugar', not a very satisfactory name, for though it does occur naturally in fruit, it is usually accompanied by glucose and sucrose. It is much sweeter than glucose, and is just as readily fermentable by yeast enzymes. Its alternative name, 'laevulose', from the Greek word *laevo*='to the left', originates from the fact that its solution rotates the plane of polarised light strongly to the left, and it is therefore curious to find that chemists write the form of fructose that occurs in nature as *d*-fructose. The explanation is that the initial '*d*' does not refer to the effect on polarised light at all, but to the arrangement of atoms in the molecule that resembles that of *d*-glucose in a certain way. This point is elucidated in the section on the chemistry of sugars.

It can, of course, be manufactured by the acid hydrolysis of sucrose, when it will be formed together with glucose, and the latter can then be separated by crystallisation, but commercially it is made by the hydrolysis of another basic starch-like substance called *inulin.* This is obtained from the tubers of certain plants, particularly dahlias, chicory (*Chicorium intybus*) and Jerusalem artichokes. This reserve polysaccharide is unusual in being a *fructosan,* breaking down to yield entirely *d*-fructose.

3. Galactose

This simple sugar, a monosaccharide, needs mentioning, because with glucose it forms the milk sugar lactose, a disaccharide. It also appears in more complex sugars still, such as raffinose, a trisaccharide, which is found in beet molasses and cotton-seed meal. It appears, too, in galactans, polysaccharides contained by certain seaweeds and lichens as gums and pectins.

When completely hydrolysed, raffinose gives equal amounts of galactose, fructose and glucose:

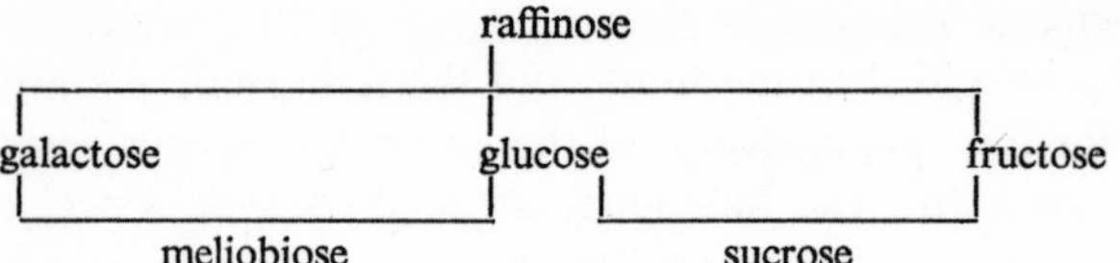

The enzyme *sucrase* splits raffinose into fructose and melibiose. As was pointed out in the section on yeasts, a feature of the lager yeast *S. carlsbergensis* is that it also contains the enzyme *melibiase*, which breaks down melibiose into galactose and glucose, so that all three simple sugars are then fermented out.

Galactose is an *isomer* of glucose, closely resembling it, except in some few physical properties.

4. Mannose

Mannose resembles glucose and galactose very closely, for all three are isomers, and all three are fermentable by the zymase complex of enzymes.

It can be obtained by the oxidation of *mannitol*, a sugar alcohol found in plants, particularly in *manna*, the dried juice of the manna-ash of southern Europe.

5. Honey

Until the manufacture of sugar, this was the sole source of sweetness. It is produced from the sugary fluid of flowers, called nectar, by the enzymes of the worker bee's digestive juices, and consists of about 33–38% of fructose and 34–42% of glucose. It is therefore basically the equimolecular mixture of glucose and fructose of invert sugar, with gums, oils, essences and minerals that bring its distinctive flavour and quality. This flavour is such that the use of honey in winemaking produces a distinct type of wine, even if it is employed only as a replacement for sugar and not on its own as the ingredient for mead.

Comparative Sweetnesses

Although there is no instrument for the accurate measurement of sweetness, there is general agreement on the following table of sweetness-rating. Sucrose is given the figure 10 as a base of comparison:

Sucrose	*Maltose*	*Lactose*	*Glucose*	*Fructose*
10·0	3·5	3·0	7·0	17·0

C. Polysaccharides or 'Multi-unit' Carbohydrates

The high polymers of simple sugars consist of very many units, indeed; of the two most common examples, starch contains several hundred, and cellulose several thousand, of such units of glucose in each molecule. The linkage of the units is not the same in each case, so that an enzyme that can break down starch has no effect on cellulose, although glucose is the basic sugar of both.

1. Starch

Starch is the typical form in which plants store reserve food, and it is present in the cells in the form of granules that are characteristic of the particular plant. They are insoluble in cold water, but when placed in hot water the granules swell and burst, and the granulose contents dissolve, leaving their starch–cellulose walls to be filtered off. If starch is heated alone it forms dextrins, and then passes on to give similar products to those produced by the distillation of wood. Acid hydrolysis also breaks it down to dextrins, and then successively to maltose and *d*-glucose. The diastase complex converts it to maltose and dextrins, but further organic hydrolysis involves maltase.

2. Cellulose

None of the polysaccharides is soluble in a strict sense because of the size of the molecule, although starch and glycogen form colloidal solutions, and this applies to cellulose. Unlike starch, it is not a storage form of carbohydrate but a constructional form, especially in plants. It forms their framework, the 'woody' part, and the non-living walls of their cells. Enzymes capable of breaking it down, called *cellulase*, are widespread in micro-organims but not in plants, except perhaps in seeds. Herbivorous animals, such as the cow, in the absence of such enzymes in their digestive systems, depend upon micro-organisms that are living in their stomach, and bacterial activity takes place during the process of chewing the cud, continuing through the whole of the gut. Even then, glucose is not produced by the micro-organisms, but the gases methane and hydrogen, with acetic, propionic and butyric acids; these acids are then stored by the cow in the form of carbohydrate and fat. Certain bacteria, such as *Clostridium dissolvens*, that inhabit stagnant water in ponds and bogs are also capable of producing cellulase enzymes which attack the cellulose of dead plants and cause methane to bubble up to the surface.

Hydrolysis by acid is not an easy matter, and production of sugar by this method is uneconomical, although cellulose is the most abundant of the polysaccharides. Sulphuric acid is the only acid to degrade it successfully, concentrated sulphuric slowly dissolving it. When boiled with this diluted acid under a pressure of 6–7 atmospheres, cellulose first toughens, and then produces *cello-dextrins*, somewhat similar to the dextrins of starch, followed by *cellobiose*, a disaccharide, and ultimately *d*-glucose.

3. Pectin

This substance (Greek *pektos*=congealed), which combines with cellulose in the formation of cell walls in fruits and vegetables, is a mixture of substances: the jelly-forming portion, which is a methyl ester of polygalacturonic acid, the polysaccharides galactan and araban, and some acetyl derivatives. The term 'ester' is used because it yields an acid and an alcohol when hydrolysed, viz. galacturonic acid and methyl alcohol; in addition, it gives some free galactose, some arabinose (a pentose sugar sometimes called 'pectin sugar') and some acetic acid.

The presence of pectin in wine may stabilise hazes and produce a cloudiness that is very difficult to clear. As an excess of alcohol causes pectin to form clots and strings of jelly, it is easy to test for it by adding three or four teaspoonfuls of methylated spirits to a spoonful of wine in a small bottle, and shaking vigorously to see whether clots form. It is this jelly-forming property of pectin that is of value in jam-making; assisted by sugar, fruit juices set after being heated to destroy their pectinesterases, and any lack of pectin for this purpose can be supplemented by commercial extracts of pectin, manufactured from *pomace*, the residual pulp from cider manufacture.

As fruit contains pectinesterase, and most yeasts polygalacturonase, pectin from ingredients used for winemaking is usually broken down naturally to soluble galacturonic acid in the fermentation vessel, but many winemakers now add commercial pectinases to their must not only to ensure clear wines but also to assist in the release of fruit juices from the cells that contain them without the need of rupturing the cells by mechanical means or by using forms of heat. Such pectinases are also of value in the after-treatment of wine that has a pectin-type haze.

4. Agar

As winemakers frequently encounter yeast cultures on agar slopes, it may be of interest to include it here, for it is a polysaccharide. It is manufactured by extraction from red seaweeds in the Far East, particularly Japan. Boiling water or hot dilute acids are used, and following neutralisation it is alternatively frozen and thawed to purify it. It is purchased usually as a greyish powder, and this is added to the culture medium and heated to boiling point. When cool, it gives a firm jelly. It appears to consist of galactans, like pectin, that are partially sulphated. We owe the idea of adding agar to nutrient broth in the place of gelatine to a certain lady, named Dr Hesse, who had encountered its employment in the Dutch Indies as a kitchen foodstuff in soups.

Chapter 7 Sugar Analysis

Solutions of substances which contain *asymmetric molecules* will rotate the plane of polarised light, and sugars are examples of such substances. This presents us with a convenient method of analysing and distinguishing sugars, and the instrument used for this purpose is called a *polariscope* or *polarimeter.*

In order to understand how this instrument works, it is necessary to realise that light is a form of wave-motion, in which the waves vibrate in up-and-down movements, something like those produced by shaking a length of cord:

These up-and-down movements occupy all possible planes perpendicular to the path of the ray of light, or, to put it another way, they continually move across the path of the ray at right angles 'all round the clock'. Thus, if one imagines a clock-face straddling a beam of light at right-angles one vibration will be up and down from the figure 12 to the figure 6, another from the figure 1 to the figure 7, another from 2 to 8, and so on, with all possible positions taken up right round the clock.

vibrations in all planes of ordinary light

If the beam of light were directed at a number of very narrow adjoining vertical slits only those vibrations which move up and down vertically would be able to pass through; if the slits were horizontal, then only vibrations moving horizontally could pass through, and so on for any position of the slits. When the vibrations of a beam of light are controlled in this way so that it continues to shine by means of vibrations on one plane only, then this light is said to be *polarised.*

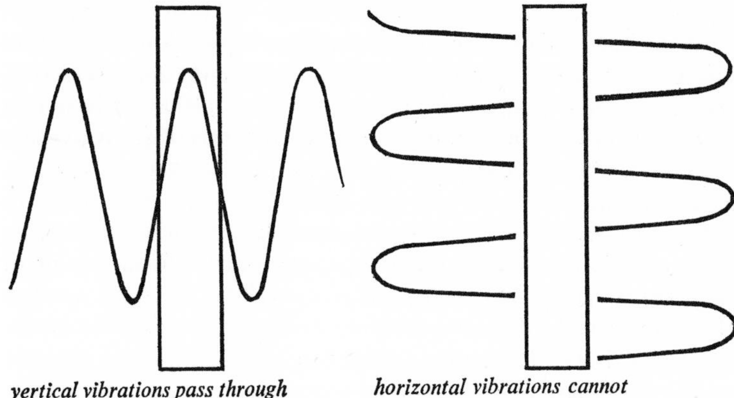

vertical vibrations pass through vertical slit

horizontal vibrations cannot pass through vertical slit

A general method of polarising light is to pass it through 'polaroid film', made of cellulose nitrate, and containing tiny crystals of iodo-quinine sulphate, or through *tourmaline*, a complex silicate, or through *Iceland spar*, a transparent and crystalline form of calcium carbonate, first brought from Iceland. If two slices of this mineral, tourmaline, were cut, placed in line and a beam of light directed at them it would be found on rotating the second slice that in one position no light would pass any further. This would be because the narrow slits in the second slice were at right angles to the slits in the first, and the polarised light emerging from the first slice could not pass any further. Further rotation until the slits in both slices became parallel would then enable the polarised light of the beam to pass through. (The term 'slit' is used to explain the phenomenon; there are no actual *slits* in a polarising filter or Iceland spar.)

The polarimeter, or *saccharimeter* as it is often called where sugars are the subject of tests, is based on this scheme of twin lenses. A beam of light is first polarised by passing it through a 'Nicol prism'. The latter consists of two wedge-shaped pieces of Iceland spar, joined together so that the plane-polarised ray passes on, while the ordinary ray is deflected. This first Nicol prism is called the *polariser*, while at the far end of the tubular instrument is a second Nicol prism, called the *analyser*, which is capable of being rotated, together with a scale marked in degrees. The polarised ray passes through the analyser, which is rotated until no light emerges from it, because the analyser is 'crossed', i.e. its slits are at right angles to the slits in the polariser.

Now a tube containing a solution of the sugar to be tested is inserted in the tube *between* the two prisms, and light emerges again from the analyser because the sugar has rotated the polarised ray. The analyser prism is now turned again until the light ceases to emerge, and the degrees of rotation necessary are carefully noted. The extent of this rotation depends upon the type of sugar (and its solution strength), and in this way various sugars are quickly distinguished.

This property of sugar to rotate clockwise or anti-clockwise the plane of polarised light is due to the arrangement of the atoms in its molecule, and though certain sugars may have the same number of the same type of atoms, different arrangements of these atoms result in a different effect on a polarised beam. Sucrose in solution rotates it strongly to the right, and it is therefore said to be *dextro-rotatory*. A glucose solution likewise is *fairly* strongly dextro-rotatory, but fructose rotates it *very* strongly to the left, and is therefore described as being *laevo-rotatory*.

One can now see the reason for the name 'invert sugar'. As sucrose is hydrolised into separate amounts of glucose and fructose, the strongly laevo-rotatory power of the fructose progressively exerts its influence over the weaker dextro-rotatory power of glucose, and by the time the sucrose is completely changed this mixture of glucose and fructose has become laevo-rotatory. In other words, the rotation of plane-polarised light has been changed from right to left, or inverted.

Chapter 8 Testing for Sugar Content

The sugar content of a liquid can be easily ascertained by means of an instrument called a hydrometer. Its use is based on a number of related effects in Physics, or more exactly that branch of it called Hydrostatics, which are governed by the law known after its discoverer as Archimedes' Principle.

Archimedes' Principle

These effects are:

1. Objects submerged, or partially submerged, in a liquid have an amount of buoyancy or 'upthrust' acting on them.

2. The amount of this upthrust on an object is equal to the weight of the water displaced by that object.

3. Consequently, the loss in weight of a submerged object is also equal to the weight of the water displaced.

Suppose that a stone is tied to, and suspended from, a spring balance, and weighs, say 320 g. We immerse it in a jar of water, and because of the upthrust it now weighs by the balance 210 g. The weight of the liquid it displaces is measured, and found to weigh 110 g—exactly the loss of weight of the submerged stone.

4. The heavier this liquid, or in more precise terms, the greater its density, the stronger the upthrust exerted.

Everyone knows how much easier it is for this reason to float in salt than in fresh water. Suppose the same stone above is immersed in salt water, which has a greater density than fresh water because of the salt dissolved in it. This time it weighs only 200 g. The amount of water displaced is the same as before, but because it is denser, or heavier, this time it weighs 120 g.

5. If an object floats on a liquid it displaces a volume that is equal in weight to itself, and the upthrust is equal to the weight of the object. This is the Law of Flotation.

To demonstrate this, suppose a small light dish, weighing 5 g is placed on water. The displacement is weighed and found to be 5 g. If a gram weight is placed in the dish it will sink lower in the water, and the extra displacement will weigh the extra gram. As this is continued by adding further weights, the tray will sink lower and lower, until eventually the weight of the tray-load will be greater than the volume of water it can displace, and it will sink.

On these interrelated facts is based the principle, discovered by the Greek, Archimedes, 287–212 B.C., that when a solid is wholly or partly immersed in a fluid the apparent loss in weight of the solid equals the weight of the fluid which it displaces. Legend has it that he discovered this when he jumped into a full bath, and that in his excitement he ran through the street in the nude, shouting, '*Eureka!*' —'I've got it!'

Density and Specific Gravity

If the weights of different liquids and solids are to be compared with one another it is necessary that equal volumes of the substances should be weighed. This is not always practicable, and the usual method is to find the weight of a *unit quantity* by dividing the weight by the size of a sample, or as it is usually expressed, by *dividing its mass by its volume*; the resulting figure is known as the *density* of the substance:

$$\text{density} = \frac{\text{mass}}{\text{volume}}$$

The units for these two measurements can be grams and cubic centimetres, or pounds and cubic feet. The units chosen must always be stated with the figure. Thus if a block of zinc $5 \times 4 \times 10$ cm weighed 1424 g, its density would be $\frac{1424\text{ g}}{200\text{ c.c.}}$ i.e. 7·12 g/c.c.

The densities of metals and liquids of different kinds have been ascertained in the laboratory, and are available when quick reference is needed. In this way an engineer dealing with huge metal structures can arrive at their weight by calculating their volume and using this in conjunction with the known density of the metal. For practical purposes it can be assumed that equal masses have equal weights:

$$\text{mass (weight)} = \text{density} \times \text{volume}$$

merely a rearrangement of the earlier formula.

Another means of comparing liquids and solids is to ascertain the ratio between their densities and the density of some substance taken as a standard of comparison. The density of distilled water at 4° C (temperature affects expansion and volume, and hence the density figure) is regarded as the standard, and the ratio between this figure and the density values of other substances is called their *Specific*

Gravity. So:

$$\text{specific gravity} = \frac{\text{density of specimen}}{\text{density of water}}$$
$$= \frac{\text{mass per unit volume of specimen}}{\text{mass per unit volume of water}}$$

and therefore specific gravity and *relative density* are the same. As this value is a *ratio*, it is always a plain figure, not expressed in units of measurement as is density, and it tells us how much denser or less dense a substance is than water. Whatever units are being used, it will have the same numerical value.

The density of water is approximately 1·00 g/c.c., that is, 1 c.c. of water weighs 1 g. Therefore the density of substances *in the metric system of units* happens to be the same as their S.G. or relative density, because to arrive at their S.G. they are divided by 1·0 in the formula above.

This is not so, however, if we use units of lb/cu. ft for their densities, for then the density of water is 62·3 lb/cu. ft, and from the S.G. formula it will be seen that their S.G.s will then be the value of their densities divided by this figure, or, to look at it another way, their densities in these units are 62·3 times their S.G. value. It is usually an advantage to use relative density or specific gravity rather than density, because they are ratios and therefore no units are required.

Some S.G.s of liquids are given for comparison:

substance	*s.g.*	*substance*	*s.g.*
Ethyl alcohol	0·789	Pure water	1·000
Kerosene	0·800	Sea water	1·030
Methylated spirits	0·830	Glycerine	1·270

The Hydrometer

By means of the hydrometer the S.G. of any liquid can be read off at a glance without the need to weigh an exact quantity in a 'density bottle' and make a calculation. A common form of this instrument resembles more or less a long glass thermometer, consisting of a long stem attached to an elongated bulb. The latter is weighted at the bottom so that when placed in the liquid being tested the instrument floats in an upright position, with the bulb immersed and part of the stem projecting above the surface. The relevant reading is where the level of the liquid cuts across the stem, and this position varies according to the liquid's density.

If the density is high the upthrust or buoyancy of the liquid will be relatively strong, the hydrometer will ride higher and more stem will project beyond the surface. If the density is low the upthrust will be relatively weak, the instrument will sink lower and less stem will project. Clearly, then, different S.G. readings will be shown in each case on the stem at the intersection of the surface. In both cases the hydrometer sinks until its weight equals the weight of the volume of liquid displaced, but this *volume* will naturally be smaller with the heavier liquid of high density than with the lighter liquid of low density: hence the hydrometer rides higher in the first instance and lower in the second.

Hydrometers came into use towards the end of the seventeenth century for ascertaining the strength of spirits. At first they were very crude, with the stem marked by immersion first in water and then in the strongest alcohol available. 'Proof spirit' was about midway between these two marks. As duties on spirits became an important source of revenue, it became necessary to use a more accurate instrument, and Customs and Excise adopted in 1816, the *Sikes hydrometer*, named after its inventor, Bartholomew Sikes. It is still the standard alcoholometer for Revenue purposes, and one often sees 'degrees proof' expressed synonymously as 'degrees Sikes'. The stem is calibrated 0–10, but the scale is arbitrary, tables being employed to give the S.G. It is made of gold-plated brass with a round bulb. Nine gold-plated brass weights accompany it, and their purpose is to extend its range, so that one instrument with an uncrowded scale can still cover a wide range of spirit values.

The French equivalent is the *Gay-Lussac hydrometer*, named after a famous French physicist, 1778–1850, with its scale representing the percentage of alcohol by volume, still the normal method of measurement in France. It is for this reason that Cognac is sometimes marked 40°, indicating '40° Gay-Lussac' and equivalent to 70° proof in England, or '70° Sikes'.

A few more famous hydrometers are worthy of mention, as their names often occur:

The Bates Hydrometer

Essentially a saccharometer, it is used in breweries for testing the original gravity of the wort in terms of sugar content, as the duty exacted turns on this figure. Its gold-plated metal is similar to the Sikes instrument, although its stem is rectangular, as is the extension of its range by the addition of round weights below the bulb. In this way each of the thirty degrees in the scale can represent as little as 0·001 S.G., and very accurate measurement results.

The Baumé Hydrometer

This French instrument, invented in 1768, owes its popularity to the complicated scales and interpretation tables of its rivals of that period. It is still used in a number of countries, but the different values attached to its readings have led to much confusion, so that there is a need either for standardisation of these or its replacement by one superior in this respect.

The Brix Hydrometer

A saccharometer, with a scale to indicate percentage of sugar content by weight.

The Twaddle Hydrometer

Each degree of the scale represents 0·005 S.G., and as no weights are used, six glass hydrometers constitute the set, with the result that the range is not limited but the outfit rather cumbersome.

The Nicholson Hydrometer

So far the hydrometers, whatever specific purpose has been served, have all been 'constant mass' instruments, that is, the means of determining density is the variation of displacement caused by a submerged hydrometer of constant mass. Density can also be determined by the variation in mass necessary to immerse a hydrometer to produce a constant displacement. The Nicholson hydrometer is such a 'constant-volume' instrument, where a variation of mass of the instrument is possible, and the volume of liquid displaced is constant. There is a single mark on the neck, which is consequently short, and a tray on the top of the neck for the addition of weights. The instrument is made to float at the set mark by adjustment of weights according to the density of the liquid and the corresponding upthrust of the liquid. The weights used indicate the liquid's S.G.

The Hydrometer in Winemaking

Hydrometers are used for a variety of practical purposes, which are sometimes indicated by the employment of a specific name for the instrument. Thus an alcoholometer is used for grading spirits, a lactometer for the quality of milk, a salinometer for the strength of salt solution. It might, therefore, be a more correct practice to term the hydrometer as used in winemaking a *saccharometer*, since its basic purpose is to indicate the sugar content of a must, although one can then continue from this to deduce the amount of alcohol formed. A hydrometer is of no use at all for the *direct* determination of the alcohol content of a particular bottle of wine. We are not dealing here with a simple aqueous solution of alcohol, but with a complex juice containing not only acids, pigments and substances such as glycerine but also probably a varying amount of residual sugar, so that the S.G. alone of the finished wine is no guide to the amount of actual alcohol present.

In winemaking a hydrometer has three main uses:

1. It indicates directly the natural sugar content of a must prior to fermentation. Obviously the greater the amount of sugar dissolved in the juice, the denser will be the liquid containing it, and the greater density will be shown by a higher S.G. reading on the hydrometer. By reference to tables supplied in instructional books on winemaking, such readings are converted into pounds and ounces of sugar per unit volume, e.g. per gallon, and it is then a simple matter to decide to what extent supplementary sugar needs to be added in order to produce the dry, medium or sweet wine of one's choice.

2. It indicates indirectly the amount of alcohol produced during a fermentation. As the sugar is consumed by conversion into alcohol by the yeast, so the density of the must falls, and the use of the hydrometer with samples at intervals enables one to follow the course of the fermentation as the sugar is used up. Eventually when this has finished a final S.G. reading is taken of the completed wine and subtracted from the original S.G. figure taken at the commencement. By reference to tables, this drop is then converted into percentage content of alcohol. In arriving at the amount of alcohol formed, notice, first, that the hydrometer indicates the amount of *sugar* consumed by the yeast as a fall in S.G. of the must; it does not measure the alcohol content *as such*, and this must be estimated by its relation to the quantity of sugar that has been converted. Secondly, the original or starting S.G. of the must is necessary as well as the final figure when the fermentation is completed; one reading is of no use without the other in ascertaining *actual* alcohol content, although it is true that the original S.G. figure alone is a guide to the *potential* alcohol content, that is, the amount of alcohol that will be produced if the sugar is fermented right out.

A drop of '100' (e.g. 1·110 to 1·010) on the hydrometer scale indicates that about 13·5% v/v of alcohol has been formed, and therefore a useful method of arriving directly at the alcohol content of a finished wine without the aid of tables is to divide the difference between first and last hydrometer readings by 7·36. In the absence of accurate tables, this gives a satisfactory estimate for home winemakers.

3. It indicates whether a wine is medium, dry or sweet. This test can be applied to any finished wine, irrespective of readings taken

at earlier stages of its fermentation, because we are ascertaining here its final sugar and not its alcohol content. It is often helpful to know this, although not an altogether satisfactory test. First, there is not complete agreement on the S.G. figure that distinguishes particular types of wine, but probably sweet wines should be over 1·020, medium wines around 1·010 and dry wines 1·000, while really dry wines should be below this last figure. Secondly, since the density of alcohol is below that of water, as in the table given earlier, its presence lowers the hydrometer reading, so that there is probably rather more sugar in the wine than the final figure suggests. It is for this reason that a really dry wine should produce a lower reading than 1·000, the S.G. of pure water. Incidentally, suspended solids can likewise interfere with the accuracy of original gravity readings at the start of the fermentation. Thirdly, the final criterion of a dry, medium or sweet wine is the palate, and it is curious how other constituents of a wine can mask the sugar it contains. An over-acid wine can appear to be dry to the palate while shown by the hydrometer to contain an amount of sugar, and as wine is meant to be drunk and not analysed on the bench, the palate is finally right.

The hydrometer is not therefore an entirely accurate instrument in winemaking, but since the alternative would seem to be an expensive instrument such as the ebullioscope, on the one hand, and rule of thumb, on the other, all in all the hydrometer, used wisely and with a sense of its weaknesses, can be of practical assistance in the control of quality winemaking.

Section 3 Alcohols

Chapter 9 The Uses of Alcohol

A Potable Alcohol

The name 'alcohol' came to Europe from the East. It is a word of Arabic origin, and is composed of two parts, the article *al* plus the term *kuh'l* (*kahala* = to stain), with the meaning 'the powder', a term applied to fine powder used by Eastern beauties for painting their eyebrows and eyelashes. As the word was accepted by European languages, it designated for a considerable period of time any fine impalpable powder. By the time of Paracelsus the last of the alchemists, who lived 1493–1541, we find the term beginning to be used to denote a volatile liquid, and in the writings of this famous personality we come across the phrase 'alcool vini', with the meaning 'alcohol of wine'.

There are very many different kinds of alcohol, but when the term is used loosely as by winemakers, it inevitably applies to the potable alcohol called *ethyl* alcohol, or *ethanol*, the common ingredient of alcoholic drinks of all types. As a pure substance, it is a colourless liquid with an agreeable odour that is often described as 'ethereal', a description no winemaker would dispute. It mixes easily with water in any proportion, and where quantities are mixed there is a contraction in volume, so that if, say, 2 gallons of alcohol are combined with 1 gallon of water the result will not be the full 3 gallons expected, and the loss is not to be attributed to the personal failings of the mixer. It has an affection for water, so that it will absorb moisture from the air if allowed to do so, and the same propensity accounts for the burning effect on the tongue when neat alcohol is being tasted. Alcohol evaporates, and even H.M. Customs and Excise make allowance for this; while Scotch is maturing under bond the distiller is allowed 2% evaporation per annum, plus 2 gallons for hogsheads or 3 gallons for other casks. This might suggest that wine maturing in cask decreases in strength because of alcohol loss, but this is not necessarily the case, for water evaporates too. The smaller water molecule moves through the wood at a faster rate than that of the alcohol, but in fact its rate of evaporation turns on the humidity of the air surrounding it. In the damp atmosphere of a cellar the evaporation of water will be slower than that of alcohol, which is unaffected, and the strength of the wine will decrease proportionately, whereas in a dry cellar the reverse will be true. Alcohol has a low boiling point, 78·4 C, compared with water, and a difference between b.p.s of two volatile liquids enables the process of distillation to take place. It burns easily in air, so that oxidation is possible, and then gives a blue, smokeless flame, producing water and carbon dioxide.

Regarded now from a medical point of view, alcohol is a chemical, a food and a drug. As a chemical, it has certain well-defined physical and chemical properties, some of which have just been mentioned; its chemical structure is dealt with in detail in a later section. As a food, it has definite calorific value, for alcohol contains a large amount of energy, as can be seen easily enough by the heat evolved when it burns, and by the fact that it is able to be used as a source of industrial power. There are roughly 200 calories in one ounce of alcohol, so that a bottle of wine containing, say, two and a half ounces of alcohol can supply some 500 calories, equivalent to a steak weighing well over a pound. In making these comparisons with food, however, one should be careful not to think too closely in terms of nutrition, and it is better to speak of alcohol's energy-value rather than its food-value. The reason is that although in its favour is its very easy and fast assimilation by the body, the same characteristic also results in its rapid elimination, and there is no way in which its energy-value can be retained and stored in the body, to be drawn upon when required, as with carbohydrates. The process of oxidation or energy-extraction starts at once, and in about two hours one ounce of alcohol will have been eliminated from the blood. The liver oxidises the alcohol along the path followed by the *Acetobacter*, the vinegar bacteria, viz. acetaldehyde–acetic acid–carbon dioxide and water, the latter end-products passing out through the lungs and kidneys respectively. Some is excreted in the urine without metabolism. These comments are restricted to the alcohol content of wine; sweet wines, of course, contain in addition carbohydrates, and there are mineral salts and vitamins in trace amounts, but important from the aspect of nutrition.

It is interesting to note that our Government in its wisdom quickly decided that alcohol is 'not a drug within the meaning of the Act' when the National Health Service Act was passed, so that doctors are unable to prescribe it for their patients or themselves at the Government's expense. Nevertheless, politics apart, its function as a drug is seen in general in its effects on the organs of the body and on the central nervous system. Here it can be regarded from many commendable points of view: an analgesic for pain, a sedative and tranquilliser for unrest, a vasodilator for high blood pressure—a tot of spirits has often saved life in an acute anginal attack, a diuretic and stimulant for the kidneys, a tonic for the convalescent. Basically, its supreme value lies in its sedative powers, its ability to release tension, remove inhibitions and to assist the individual to 'come out of his shell' and pass beyond the petty frictions of everyday life to the state of bonhommie that poets and musicians have recorded down the centuries. Sometimes reference is made to alcohol as a stimulant, but any stimulus that follows a drink is really the result of the ensuing relaxation that permits the emergence of personality, coupled with easily available energy, so that it is more exact to regard alcohol as a sedative than as a true stimulant to the central nervous system, despite the sense of stimulation that is felt.

If a person persists in drinking he passes on through various stages of merriment until the nervous centres controlling in turn sight, speech and gait are affected, and he becomes a drunken nuisance, finally ending in stupour and coma, which at a blood-alcohol level of about 0·6% is diagnosed as alcoholic poisoning. Alcohol is not alone in changing from a friend to an enemy when it is abused. In the right amounts—and there may be some difference of opinion here!—it can cheer; in excess it can injure; used insanely it can kill. Samuel Butler said, 'No one can hate drunkenness more than I do, but I am confident the human intellect owes its superiority over that of the lower animals in good measure to the stimulus which alcohol has given to the imagination.'

At the time of writing the Government has accepted the B.M.A.s recommendation of an arbitrary figure of blood-alcohol level which automatically makes any driver with that alcohol content liable to further tests of blood and urine with a view to a drunken driving charge. The standard of measurement is in milligrams of alcohol per 100 millilitres of blood, and the guilt-limit has been fixed at 80 mg. The culpable level of alcohol varies considerably in different countries, Germany and Belgium regarding 150 mg/100 ml as evidence of intoxication, whereas with Sweden and Norway the figure is 50 mg/100 ml in compulsory tests. Apart from anything like universal agreement on what is the correct figure of blood-alcohol concentration, any automatic guilt-limit applicable to all and sundry is open to abuse, for it overrides individual variation of reaction as well as environmental conditions of drinking. The extremely experienced driver who uses a car for long periods each day may conceivably constitute less risk on the road with 80 mg of alcohol per 100 ml of blood than the sober novice new to the road who has not been within sight of a bottle. Then, too, the hardened and regular drinker has a nervous system far more tolerant of alcohol than the occasional drinker out on the spree, so that the latter could be incapable at a level where the former still shows little response to what he has drunk. Another factor that is ignored is the physical condition of the drinker at the time. Not only are there these variations between individuals but the same person is affected by certain circumstances in different degrees, and the type of drink, its nature, concentration of ethyl alcohol and percentage of methyl and fusel oil is one of these.

The amount of water combined with the alcohol considerably affects the absorption rate. As wine-drinkers in this country do not follow the habit in France of 'cutting' wine with water, the beer-drinker is at an advantage here, for he will be moving surely, albeit waveringly, towards home and beauty while his wine-drinking colleague, having consumed the same amount of alcohol but in a less diluted form, is still resting under the table. Food taken with drink is also a circumstance that slows down the absorption rate considerably. Not only does the presence of food mop-up the alcohol and reduce its contract-area in relation to the lining of the stomach but the steady movement that takes place internally retards the passing of the alcohol into the small intestine. The presence of alcohol in this part of the bowel makes itself known with extreme rapidity, and anything that can hold back this is very desirable. The famous wine connoisseur, George Saintsbury, denied this point, but he was a curiously obtuse man in some ways, and certainly incorrect on this matter. Those who anticipate a round of drinks without the assistance of food might like to follow the advice of that seasoned drinker, Maurice Healy, who recalls the old Roman practice, in his book *Stay Me with Flagons*, of taking a couple of spoonfuls of olive oil prior to drinking: 'When alcohol reaches an oily stomach, it cannot immediately get at the nerves.' In the author's opinion raw egg and milk is more efficacious still.

As regards drink and driving, one must come to accept the fact that if a blood-alcohol level of 80 mg/100 ml is not a convincing measure of drunkenness, at least it measures an average level of impairment of driving efficiency. Reference to a relevant graph will make this clear, although everyone will observe that the figures given are averaged, and everyone will know for certain that his own personal performance is far superior to the norm!

B. Industrial Alcohol

In addition to its potable value, alcohol is of the greatest importance in two further ways: as a source of organic energy for heat, light and motion, and as a raw material for industrial manufacture.

As a source of power, alcohol is usually combined with other fuels in motor engines, for when used by itself it needs a higher compression ratio than is suitable for this type of engine, and starting from cold also presents problems that are not easy to overcome without special modifications. Together with ordinary petrol, in the proportion of about half and half, it performs well, and the German national fuel called 'Monopolin', composed of 45% alcohol, 45% petrol and 10% benzol, is an example of such a practical mixture. The main possibilities of alcohol in this direction, however, seem to lie in the tremendous surge of rocket power that can be produced when alcohol and liquid oxygen are the basis of combustion, as they were in the German V2's that harried the United Kingdom during the last war.

As a raw material, alcohol is of value in the manufacture of an endless array of goods: dyes, inks, plastics, textiles, paints, medicines, anaesthetics, synthetic rubbers, chemicals—even felt hats! Naturally, a very tight control indeed is kept on alcohol intended for these purposes, and means are devised to save employees who handle it in the course of their daily work from the strain of temptation. All alcohol intended for industry or for locomotion is kept under bond by licensed methylators, who then mix it with the listed substances called *denaturants*, under the eye of Customs and Revenue officers. In this way the payment and repayment of large sums of money in duty are obviated.

There are four types of such denatured alcohol, known as *methylated spirits*, dependent on the general purpose for which it is intended:

1. Mineralised methylated spirits.
2. Industrial methylated spirits.
3. Industrial pyridinised methylated spirits.
4. Power methylated spirits.

It will suffice to describe the first type, which alone is allowed to be sold to the general public, and then only by licensed retailers in quantities limited to 4 gallons at any one time. It consists of 90 parts ethyl alcohol, 9½ parts methyl alcohol and ½ part crude pyridine. Methyl alcohol, CH_3OH, sometimes called *wood spirit* because it was formerly obtained by the destructive distillation of wood, is toxic when taken in any quantity, and of course it gives its name to the denatured spirit. It is not possible to separate these two alcohols by any home-distilling process, as the methylators are well aware. In order to discourage those whose palate is not averse to the presence of methyl, a substance called *pyridine*, C_5H_5N, is added. When the dry distillation of bones takes place *bone-oil* is formed, a most foul, sickly smelling dark liquid. From this, the colourless liquid pyridine is produced, with its pungent bouquet of burnt feathers. The more usual source of pyridine today is coal tar. The methylators add this, together with a dash of petroleum oil and methyl violet dye.

Chapter 10 Types of Alcohol

A. Methyl Alcohol, CH_3OH, b.p. 65° C

This is the simplest of all the series of alcohols, resembling ethyl alcohol in its smell and appearance, although perhaps rather more pungent, with a lower boiling point of 65° C. Formerly, it was prepared exclusively by the *destructive distillation* of wood, a term that implies the extraction of substances from solids by their decomposition. It was the famous English chemist Boyle, in 1661, who first noted the presence of wood spirit in the distillation, and the Frenchmen Dumas and Péligot named it 'Methyl alcohol' (*methanol* is the modern term), from the Greek *Methu*=wine and *hule*=wood, in 1834. The wood is put into iron containers and heated until these are red hot, with the result that the complex carbohydrate compounds break down. Wood consists mainly of two ingredients, cellulose fibres, made of huge polysaccharide molecules, and *lignin*, which acts as a binder for the fibres, and contains methoxyl groups and benzene rings. From the fumes produced, tar and a considerable amount of watery liquid is condensed, the latter being known as *pyroligneous acid*, from Greek *pȳr*=fire, and Latin *lignum*=wood. This distillate contains an extraordinary variety of substances, but from our point of view we need notice only acetic acid and methyl alcohol. The wood-distilling industry is still alive, but its most active period was 1850–1920.

Products from Wood-distillation

Gas: methane gas, carbon monoxide and carbon dioxide.

Pyroligneous acid: methyl alcohol, acetic acid and acetone.

Wood tar: an oily, insoluble liquid.

Charcoal, the unconsumed solid residue.

Methyl alcohol cannot be prepared by fermentation as with ethyl alcohol. Its presence in wine arises from the breakdown of pectin by the enzyme pectinesterase to give pectic acid and a small amount of methyl alcohol. It is made synthetically by passing a mixture of water gas and hydrogen over a catalyst of chromic zinc oxides at 200 atmospheres pressure and at a temperature of 450° C. The basic reaction is:

$$\underset{\textit{carbon monoxide}}{CO} + \underset{\textit{hydrogen}}{2H_2} \longrightarrow \underset{\textit{methyl alcohol}}{CH_3OH}$$

Water gas is a mixture of carbon monoxide and hydrogen, made by passing steam over white-hot coke, and thus it has coal as its origin. Of recent years the use of 'synthetic gas' has increased as the basis for methyl alcohol production. This is the term employed for a mixture of carbon monoxide and hydrogen that derives from petroleum sources; in the United States 99% of the methyl alcohol production originates in this way.

Methyl alcohol is of great importance to industry; it is used in the synthetic preparation of chemicals, it is used as a solvent for varnishes, it is the source of *formaldehyde*, used as a disinfectant and for plastic manufacture, and of course, sadly enough, it is used to denature ethyl alcohol. In any quantity it is toxic when consumed by the body, and from time to time one reads of deaths from this cause, although the trace amounts in wine are harmless enough and contribute to the bouquet and vinosity.

B. Ethyl Alcohol, C_2H_5OH, b.p. 78° C

The molecule of ethyl alcohol is a little more complex, as its formula shows, but it is only second in the huge family of alcohols and still a relatively simple example. Together with methyl alcohol, it represents a typical 'primary alcohol'. This, of course, is the alcohol that is produced by the zymase complex of enzymes in yeast, and is the basis of beers, wines and spirits. Its qualities have been discussed, and as its chemical properties and its manufacture are dealt with in later pages, there is no need to repeat the information here.

C. Higher Alcohols

This term is in general use for the alcohols of higher molecular weight than methyl and ethyl alcohol, i.e. those which have more than two carbon atoms in their molecules. It will have been noticed that methyl alcohol has *one* carbon: CH_3OH, and ethyl alcohol *two* carbons: C_2H_5OH. There is a very large series of alcohols, the formula for each increasing by an additional $-CH_2$ increment as the series ascends; thus, the third in the series, *propyl alcohol*, has the formula C_3H_7OH, *butyl alcohol* C_4H_9OH, *amyl alcohol* $C_5H_{11}OH$ and so on. Notice the –OH of each formula, an oxygen and hydrogen atom linked together, and called the *hydroxyl group*, which is the distinctive group of *any* alcohol, wherever it occurs in the series.

Throughout the series the chemical properties remain similar, but the physical properties vary regularly as the number of carbons increases. Thus, the solubility in water steadily decreases as the carbons become more numerous, until alcohols containing over sixteen carbons become crystalline, odourless and tasteless solids, resembling the appearance of waxy solids. Likewise, the boiling points of alcohols increase regularly by about 20° C for each increment of $-CH_2$: *n*-propyl boils at 97° C, *n*-butyl at 117° C and *n*-amyl at 137° C and so on. *Secondary* and *tertiary* alcohols have lower boiling points respectively than their related primaries.

Of the large number of higher alcohols possible, only the three mentioned above need interest the winemaker, as occurring in his wine in trace amounts. This simplifies matters considerably, but even so, a complication arises from the fact that each of these higher alcohols occurs in a number of different forms, a situation that at first is very puzzling. The cause is the occurrence of isomers, a phenomenon that we encountered in sugars. Alcohols can have immense numbers of isomers, because not only do those with the same molecular formula vary in their structural arrangement but also according to where the hydroxyl group is attached. This is clarified in the last section on the chemistry of alcohols.

The net result is that there are two propyl alcohols, four butyl alcohols and eight amyl alcohols. This is the reason for such terms as *n*-propyl alcohol, meaning *normal* propyl alcohol, and *iso*propyl alcohol, meaning *isomer* propyl alcohol, which has the same molecular composition as the normal type but a different structural arrangement. As the $-CH_2$ increments increase, so the number of isomers increases too, until they run into thousands for each alcohol. The position of the hydroxyl group in the molecular arrangement gives a further means of distinguishing the alcohols as primary, secondary and tertiary.

The propyl, butyl and amyl alcohols are volatile, colourless and fragrant liquids, which in trace amounts bring flavour and bouquet with them to the wine. This is well known, but another property which they possess and which is not yet fully recognised, although Thorne has demonstrated it, is their power of inhibiting the yeast from carrying out the fermentation of sugar. Ethyl alcohol gradually restricts the activity of the yeast as it is formed, and this inhibition is supplemented by the higher alcohols to a degree quite disproportionate to their quantity. In a normal fermentation, however, the winemaker is unlikely to be troubled in this way.

The petroleum chemical industry plays a large part in the production of higher alcohols by synthetic means. Thus *iso*propyl alcohol is obtained from propylene, and *n*-butyl and *iso*butyl alcohol from the same source by what is termed the 'Oxo reaction'. Some can also be produced by bacteria from carbohydrates, and the so-called 'A–B fermentation' that produces acetone and butyl alcohol from 'black-strap' molasses by means of the bacterium *Clostridium acetobutylicum* is a commercial success. It has been said that wines using a 'natural fermentation' are likely to contain twice or three times the quantity of higher alcohols than wines produced by the inoculation of a purified must with a selected yeast culture, and the uncontrolled deamination of undesirable protein material, together with the risk of bacterial infection along the lines mentioned, lends support to this opinion. Grandma's potato wine was often a deadlier brew than she thought.

Fusel Oil

This is the collective name given to the mixture of inactive *iso*amyl alcohol and active amyl alcohol, which usually occurs in trace amounts in wines. The term 'fusel' is from the German word for 'bad spirit'; the expression 'foozle', known to golfers, is probably from the same stem (German dialect *fuseln* = work badly or slowly). The use of 'active' and 'inactive' with amyl alcohols assists in distinguishing them, since three of the isomers can occur in optically active forms, rotating the plane of polarised light to the left or right. In addition to these two main constituents, some *iso*butyl alcohol, a more volatile substance, is likely to be present.

Although in minute amounts fusel oil can contribute to the flavour and bouquet in wine, in excess it can cause severe headache accompanied by thirst, and in appreciable quantities it would be classified as a convulsive poison. Its occurrence in wine arises from the assimilation of amino acids by yeast enzymes, and thus although it appears *during* the fermentation process, it is neither a direct nor an indirect product of the sugar to alcohol route. The deamination of the amino acids *leucine* and *isoleucine*, followed by decarboxylation, in particular increases the amount of fusel oil, and therefore where the ingredients are known to bring high proportions of these acids to the must, such as grain, beetroots and potatoes, and incidentally the 'marc' of grapes, then it is advisable to introduce ammonium salts in order that the yeast has ample alternative supplies of nitrogen and therefore has less need to break down these amino acids into ammonia and fusel oil.

D. Glycerine, $C_3H_5(OH)_3$, b.p. 290° C

Known chemically as *glycerol,* this is classed as an alcohol, and as a constituent of wine it is included here. We have seen that the hydroxyl group, –OH, is distinctive of alcohols; some alcohols, unlike any we have so far met, contain two or more of these hydroxyl groups in their molecule. Glycerine contains three such groups, the $(OH)_3$ in its formula, and is therefore a *trihydric alcohol.*

It is a viscous, colourless, odourless liquid, with a sweet taste. Solubility and sweetness are associated with *polyhydroxy-substances,* including sugars, and the three hydroxyl groups are also responsible for the high boiling point. Glycerine mixes well with water and alcohol in any proportion, and attracts moisture from the air.

Formerly, it was obtained as a by-product of soap manufacture, from the spent lyes concentrated by evaporation in vacuum pans. For chemical use, it was then distilled. Since 1948 it has been produced synthetically from *propylene,* a major achievement of the petroleum chemical field. The result of this has been some stability in price, once very unsettled.

Of more interest to winemakers is its formation as a product of fermentation. Pasteur had shown in 1859 that glycerine is found in small amounts during the conversion of sugar to alcohol. In the normal way it occurs to the extent of about 0·5% w/v, and this can be raised to a maximum of about 3% by various means. Because of its value in the manufacture of dynamite, German scientists in the First World War discovered that by diverting sugar from its normal fermentation route by environmental changes, they could produce glycerine at the expense of alcohol. In America they also changed the metabolism of the yeast by manipulating the pH of the must, and fermenting in an alkaline solution, while in Britain the same effect was produced by adding sodium sulphite and bisulphite to the medium, a process which also inhibits bacterial infection. The yield of glycerine thus produced was about 25% of the sugar consumed, as compared with about 3% in a normal fermentation, but the process is uneconomic, especially in the face of synthetic competition.

The addition of a small amount of glycerine to a thin and harsh wine, particularly if it is over-astringent, has a remarkable smoothing effect. It is a practice frowned on by purists, but those addicted to the habit argue that as an alcohol and a natural product of fermentation its employment as an additive is no more a deviation than the fortification of a sherry with the alcohol that it cannot produce naturally in sufficient quantity for this type of wine.

An example of another polyhydric alcohol is *mannitol*, that has six hydroxyl groups: $C_6H_8(OH)_6$. It is a plant alcohol, found in juices such as of the manna-ash. It is sweet and useful as a mild aperient. When oxidised it produces the sugars mannose and fructose. *Sorbitol* is an optical isomer of this.

Chapter 11 The Manufacture of Ethyl Alcohol

This chapter is concerned with the production of ethyl alcohol beyond the percentage that occurs in any natural form of yeast fermentation, and used either as spirits for drinking or as industrial alcohol for commercial use.

A normal table wine contains 9–12% v/v of alcohol, and home winemakers, who are not restricted by natural grape sugar but chaptalise the must according to the type of wine required, are able to reach 16% of alcohol if they are so inclined. Under certain conditions yeast can be persuaded to produce even higher amounts, but despite extravagant claims made occasionally, anything beyond 18% v/v of alcohol by home winemakers can be regarded as based on a miscalculation or inaccurate reading. Saké, or rice wine, made by Japanese methods and utilising moulds in conjunction with yeast, is a special case, and this reaches 20–22% of alcohol before being diluted for bottling. It is true that research in crossing spores of yeasts to produce hybrids is proceeding, and it is known, too, that yeast strains may be acclimatised to the presence of high amounts of alcohol, so that by careful 'training' and selection they may become increasingly alcohol tolerant, but it is unlikely that such types will come easily into the hands of home winemakers for a very long time.

Wines that contain higher amounts of alcohol than this, such as sherry, port and madeira, which have about 20% of alcohol, are *fortified* wines, i.e. they have had brandy, or perhaps more exactly *eau de vie de vin*, added to increase the alcohol content. To produce higher proportions of alcohol than are possible by yeast fermentation alone, two courses are available: (1) fermentation followed by distillation; (2) synthetic conversion of hydrocarbons. These will now be considered in detail.

A. Fermentation and Distillation Processes

When the raw material is of a farinaceous nature three distinct stages are to be distinguished, the first of which is unnecessary when a sugar solution or molasses is used.

1. Starch Saccharification

The two types of spirit that do not use starch as their basic ingredient are brandy and rum. Brandy is distilled from grape-wine, the alcohol arising from the fermented grape-sugar, and the various fruit-brandies are likewise based on the natural sugars they contribute. Rum is made from fermented molasses. Apart from these two famous exceptions, the other potable spirits, namely whisky, gin and vodka, all use starch as their basic raw material.

Barley is recognised as the starch-source for whisky, particularly the strong-flavoured Scottish malt whisky from pot stills, but for general consumption outside Scotland this is blended with the lighter, less-flavoursome grain whisky, made from various combinations of grain, such as maize, barley, oats and rye, and distilled in patent stills. Gin is a refined spirit that takes its flavour from secret herbal additions, in which juniper plays a large part, so that it can be made from any form of starch. The best types, however, rely on grain mixtures selected by individual manufacturers, especially maize and rye. Vodka is similar in being refined to the extent of possessing no distinctive flavour of its own, and its resulting wide choice of raw materials include potatoes and starch vegetables, although alcohol from potato starch is considered to retain certain 'overtones'.

Industrial alcohol, when manufactured by fermentation, can also make use of any source of carbohydrates, whether sugar, grain or vegetable. Even sawdust, which exists in large quantities in Sweden, and the spent sulphite liquors from wood-pulp manufactures, can be turned to economic use in suitable localities, but not in this country. Fermented industrial alcohol in Britain relies largely on cane molasses or 'blackstrap', imported from Cuba and the West Indies, and to a lesser degree on molasses from sugar refineries. (Beet molasses are more in demand for the production of bakery yeast, since it contains less gum and cellulose than cane, the diffusion extraction of its sugar being responsible for this clarity.) Grain is extensively used in the United States, but not in this country for the last forty to fifty years.

When the raw materials selected supply carbohydrate in the form of starch, then the first stage in the production of alcohol is the saccharification of this medium. There are three methods available, two biological and one chemical.

1. *The Use of Malts*

This process relies on the fact that malted grain, and one naturally thinks of barley in this respect, contains enzymes of the *diastase complex*, called *α-amylase* and *β-amylase*, which convert starch into dextrins and maltose. The enzyme *ptyalin* of human saliva has the same effect.

The first of these, α-amylase, breaks starch down into gummy, soluble compounds called *dextrins*, acting best at a temperature of 150° F and a pH of 5·7. The β-amylase breaks down some of these dextrins into *maltose*, and also attacks the starch molecules directly, changing a proportion into malt sugar, its optimum temperature being 130° F and its pH 4·7. So, by varying the temperature and the acid content of the wort, the formation of dextrins or of maltose may be emphasised. Distillers want the maximum of malt sugar for alcoholic strength, but brewers aim at an amount of dextrins beyond the minimum attainable because they bring character and a stable foam to the beer. Starch is composed of two constituents, named *amylose* and *amylopectin*, varying in proportion according to the origin of the starch. α-amylase is able to convert both of these to dextrins; β-amylase is able to convert the whole of amylose to maltose, but only about half of the amylopectin, the remainder being left as unfermentable dextrins. The following much simplified diagram summarises these results:

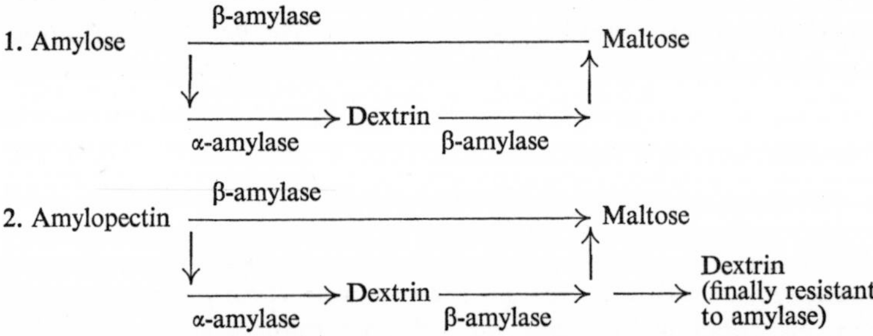

Other enzymes are also contained in the grain that attack cell walls, and *proteases* that are able to break down the protein content. All these and others are activated by the process of malting, the main stages of which we can now consider.

MALTING. The first step is to steep the barley—and we shall take this as a typical grain for the process—in water cisterns for 48–72 hours, where it takes in about its own weight in water. This is followed by 'flooring', consisting of spreading the grain over concrete malting floors in heaps or 'couches' to germinate for eight to twelve days. The rootlet or 'radicle' has then appeared, and the shoot of the embryonic plant, known as the 'acrospire', although still covered by the outer skin of the grain, can be seen pushing along its back. From time to time the barley is turned with wooden shovels or ploughs, not only to even the growth but also to aerate the grain, for air is essential to germination. This period of pseudo-germination persuades the seed that spring is here, and the enzymes are activated into their action stations, the diastase group commencing to break down the starch into sugar for the consumption of the plant-to-be.

KILNING. A cruel disappointment lies ahead for the little plant, for the next stage is 'kilning'. This consists of checking the germination, or 'modification' as the term is, by drying the grain on perforated metal floors by currents of hot air from below. As a result, it is possible to store the malted barley until wanted without further acrospire growth and, very important, further consumption of sugar by the plantlet. It also prevents the grain from going mouldy in its moist condition. Kilning has an effect on the resultant flavour and characteristics of the beverage, if such is wanted; dark beers and stouts need a higher temperature and darker malt than pale ale, while malt whisky is said to owe much to the use of peat for kilning. Naturally, a proportion of the enzymes are destroyed by the heat, but they are less affected by dry heat than by the heat of a liquid, so that sufficient survive to continue the saccharification in the last stage. Distillers want to produce the maximum of alcohol possible, and they therefore kiln lightly, in the case of malt whisky, so that

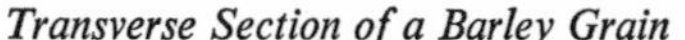

Transverse Section of a Barley Grain

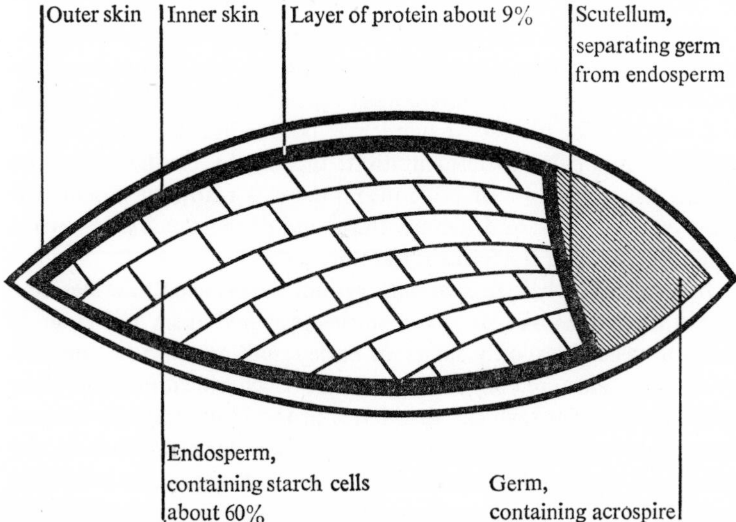

more diastatic power is available than to a brewer. Other distillers, who are not interested at all in flavour from the ingredients, such as gin distillers, may kiln even less or use a 'green malt' if convenient.

MASHING. Here the diastase enzymes complete their work started during malting. It is not always realised that although some starch was broken down to maltose in that early stage, ten to twelve times that amount is converted during the process of mashing. This consists of mixing the crushed malt, called 'grist', with water in a mixing machine called 'Steel's masher', and feeding into mash tuns fitted with mechanical stirrers, where it can steep. The water, known as 'liquor' to brewers, is heated to the temperature best suited to the α- or β-enzymes which are to be activated, the distiller preferring a lower temperature to encourage the production of maltose at the expense of dextrins. In Britain the 'infusion system' is usual, whereby the grist and liquor are mixed at the required temperature and fed into the masher, but on the Continent and in America the 'decoction system' is used, in which the mixing takes place at a lower temperature, that is then raised to the necessary figure in the tun. The latter method encourages the *proteases* or protein degrading enzymes, that enjoy a temperature of 120° F, and so provide more nitrogenous nutrient for the yeasts. Mashing not only continues and completes the malting process of saccharification but also dissolves out the maltose and dextrins, so that after about two hours in the tun no starch is left of any consequence. Thus mashing may also be regarded as the first stage in the actual brewing process.

So far, the various stages of malting, kilning and mashing are, more or less, common to the production of beer and malt spirits such as whisky, but now differences occur according to the particular end in view. The brewer we will leave, boiling his wort with hops in giant coppers prior to fermentation; the whisky distiller draws off the cooled sweet wort or wash into huge vats ready for inoculation with yeast cultures, and these with the aid of their *maltase* enzymes can now take over the fermentation process.

In distilleries where malted barley is not the principal raw material, as with grain whisky or other spirits prepared from mixed grain or from vegetables, only sufficient barley malt needs to be added to provide the diastase enzymes that ensure saccharification. This is the usual practice with whisky distilleries in the United States. In such cases the farinaceous ingredients are first cooked, often under pressure, to speed the process and 'solubilise' the starch. Starch is insoluble in water, but heating with water expands the granules and causes them to burst, so forming a colloidal solution. This is then put into mash tuns with water and the appropriate amount of barley malt agent added, the process now resembling the stage of mashing described above, where the diastase enzymes of the malt proceed to saccharify the starch.

When grain provides the raw material of starch the still considerable cost of the malt additive can be cut by utilising the enzymes contained in the ingredient itself. Although barley is usually thought of as a natural malting cereal, others are quite capable of acting as saccharifying agents. Oat malt is quickly and easily prepared, and has persistent diastatic qualities, and rye is sometimes used together with it; on its own, rye tangles badly on the malting floor. In the Balls–Tucker process wheat is ground up and mixed with water to form a porridge. Enzyme growth is encouraged by the addition of acid and sodium sulphite. After the slurry has settled the remaining liquid containing the diastase enzymes is filtered off, and used for saccharifying the grain, that has previously been cooked.

2. *The Use of Moulds*

An alternative method of converting starch to sugar uses enzymes from fungi. Some of these, indeed, can carry the fermentation of starches right through to the alcoholic stage, but in industry it is usual to employ them for dextrinisation only, and to follow with a yeast for the fermentation of the sugar formed, since in this way the alcohol yield is higher. About a hundred years ago the German, A. Fitz, demonstrated that *Mucor racemosus*, a very common mould found on damp material, can produce alcohol in anaerobic conditions.

Saké, the national Japanese rice drink, is based on the use of *Aspergillus* moulds, and J. Takamine and K. Oshima have done much research on these, comparing their dextrinising properties with malt diastase, with the result that various commercial enzymic products were put on the market for starch saccharification, such as Taka-diastase, Polyzime, Oryzyme and so on. In the Taka-diastase process the mould *Aspergillus oryzae* (from Latin *aspergillum* = 'mop for distributing holy water', a reference to its appearance, and *oryzae* = 'of rice') has its spores mixed with moist, sterilised bran, which is then placed in trays in a hot, humid temperature of about 100° F for some 40 hours or so. During this time the mould has developed and reproduced rapidly in the bran, and by means of a 20% alcohol solution its amylase enzymes are extracted and then precipitated by increasing the proportion of alcohol. They are finally filtered and dried. When the starch ingredients have been solubilised in preparation the enzymes so extracted can be added directly to the mash.

Another successful commercial process utilising moulds, which has shown its value in countries with a tropical climate where the production and storage of malt presents problems, was developed by Calmette in the 1890s, by isolating moulds containing amylase enzymes from Chinese rice. This method uses another genus, usually *Mucor* or *Rhizopus*; *Mucor rouxii* (named after the Frenchman Emile Roux) was originally used, but later various species of *Rhizopus* were preferred. In the 'Amylo process', as it is named, the mould culture is not mixed with an intervening growth medium but with the grain itself, following the initial cooking. The conversion of the starch to maltose is completed in about 24 hours, and the mash is then ready to be inoculated with yeast. An alternative method is to add both mould and yeast at the same time, as in the production of Saké, and a short description of this will serve as illustration.

Saké is one of the few wines that are prepared from starch ingredients, as distinct from beers, without extraneous sugar being added. A small amount of rice is steamed under pressure to sterilise it and to dextrinise the starch. This is mixed with spores of *Aspergillus oryzae*, and after 40–48 hours incubation the culture of the mould is complete, known now as *Koji*. With the aid of some of this saccharified rice, a yeast culture is started, known as *Moto*, and when it is fermenting vigorously portions of the koji and moto are added together to the main bulk of steamed rice. It is claimed that an alcohol content of 20–22% v/v is achieved by this method. The explanation is that without yeast the accumulation of maltose progressively inhibits the amylase action of the mould, and furthermore some starch is converted into unfermentable isomaltose. When yeast is present, together with the mould, it removes the maltose by fermenting it as soon as it is formed from the starch, and thus allows *A. oryzae* to proceed uninhibited in any way. The high alcohol content is diluted to 15–16% prior to bottling.

3. *The Use of Acids*

The hydrolysis of starch to sugar is also possible in the presence of acid which acts as a catalyst in the process. The material is ground or steeped and heated in water with a dilute mineral acid under pressure. Hydrochloric acid is usual, but sulphuric acid is sometimes used. As the resultant liquid is acid, it is necessary to neutralise its excess acidity with an alkali before fermentation can take place. A variety of suitable substances exist for the purpose, such as caustic soda, chalk (calcium carbonate), washing soda (sodium carbonate) and ammonia, but the last is particularly appropriate where the hydrolysis of starch is ultimately for the production of alcohol and not glucose, as the ammonium salts produced by the reaction of acid and ammonia serve the same purpose as 'nutrient tablets' in home-produced must, viz. the supply of nitrogen for yeast nourishment during the ensuing fermentation.

If the starch is only partially hydrolised a mixture of glucose and maltose is formed with dextrins, but where the process is completed, entire conversion to *d*-glucose is obtained. This is a stage further than is possible with diastase as the organic catalyst, which stops at maltose, and leaves further conversion to the maltase enzymes of the yeast. The acid hydrolysis of starch takes place in several stages, first forming a soluble form of starch, then breaking down to dextrins, which are then hydrolised to the disaccharide maltose, that finally breaks down again to the monosaccharide glucose.

$$\underset{starch}{(C_6H_{10}O_5)_n} \rightarrow \text{soluble starch} \rightarrow \text{dextrins} \xrightarrow{H_2O} \underset{maltose}{C_{12}H_{22}O_{11}} \xrightarrow{H_2O} \underset{d\text{-}glucose}{C_6H_{12}O_6} + C_6H_{12}O_6$$

Where organic catalysts, such as malt, are used, the conversion not only stops short at the maltose stage but also includes a proportion of unfermentable dextrins. A number of these gummy-substances exist, corresponding to different linkages in the starch-molecule chain. The largest is *Erythrodextrin,* which still gives a blue reaction, indicative of starch, with iodine. *Achroodextrin* and *maltodextrin* represent successive shortenings of the chain on the way to maltose.

In the production of 'grain whisky', that is, alcohol where a spirit without pronounced flavour is satisfactory, later to be blended with malt whisky to lighten the latter for general consumption, saccharification by acid hydrolysis may be combined with saccharification by malt additive. This practice is common in this country but not America. Before the saccharification is quite completed the sulphuric acid is neutralised by milk of lime and powdered chalk, and removed as calcium sulphate. Malt is then added, and this completes the conversion to sugar, so that a combination of acid and enzymic hydrolysis has been employed. Rectification by patent still follows as described later.

WOOD. Reference was made earlier to the use of sawdust as a raw material for industrial alcohol, and it is true that wood is another substance that may be saccharified by acid hydrolysis. This is because in addition to the tough woody substance called *lignin*, which acts as a sort of filler or binder, it contains large amounts of cellulose fibres. Cellulose consists of huge molecules of glucose units linked through an oxygen atom $(-C_6H_{10}O_5-)_n$, and its acid hydrolysis is possible, though difficult and not an economic proposition. The 'Bergius method' of treating wood to obtain sugar dissolves the cellulose part of the wood by a 40% hydrochloric acid solution, leaving the lignin behind. To remove most of the acid, the acid–sugar solution is then distilled, and the sugars recovered by evaporation. The constituent sugars depend upon the wood used as the raw material, and they present various difficulties for the chemist. Not all the sugars, the pentose wood-sugar *xylose*, for example, are fermentable, and substances are also present that inhibit yeast activity. There are ways of overcoming these problems, but they raise the cost of production.

Likewise, spent sulphite liquor can be used as a source of sugars for industrial alcohol. Wood pulp that is intended for making paper is made by boiling wood chips with calcium bisulphite under pressure. About 2% of fermentable sugars is contained in the waste sulphite liquor, and alcohol is being produced from this source of wood sugar in Scandinavia.

2. Sugar Fermentation

At this stage, whatever the raw material, the ingredient is presented in solution as sugar for the yeast to ferment. Where the material is farinaceous, such as grain and vegetable, the starch has now been converted to sugar by one of the methods outlined, and in other cases, such as grapes, fruit or molasses, the carbohydrate is already contained as a form of natural sugar.

It is an axiom, known as *Kluyver's Law*, after A. J. Kluyver, who postulated it, that any species of yeast that ferments any other kind of sugar will also ferment glucose, fructose and mannose. The disaccharides maltose and sucrose depend for their fermentation on the possession or not of the appropriate enzymes in the yeast used, for they need to be broken down into simple hexose units of glucose or glucose and fructose respectively prior to fermentation proper. Such hydrolytic enzymes are common in *Saccharomyces* species of yeasts, and therefore none of these sugars presents any particular fermentation problems from this angle.

The fermentation process is very similar for all sugars, such variations as exist being determined by the end in view. Thus the distiller obviously aims at obtaining as much alcohol as possible, whereas to the brewer alcohol is only one of many other considerations. The malt-whisky distiller, who uses a pot still, will recognise that much of the flavour of the liquor will be carried over to the final product, whereas the gin distiller is aware that his patent or Coffey still can efficiently remove all traces of its origin from the refined spirit he is manufacturing, and his attitude to the culture medium is frankly utilitarian. He therefore has no hesitation in using sulphuric acid for lowering the pH where necessary, a practice that makes the winemaker shudder!

Bearing in mind that such variations in technique exist according to the end product, we can follow the stage of fermentation as it takes place in the manufacture of industrial alcohol. The material used for this in Great Britain is usually cane molasses, either imported direct or from local sugar refineries, and therefore, of course, no saccharification has been necessary. Of the 50% or so sugar content of the molasses, about two-thirds consist of sucrose and one-third of invert sugar. The molasses is diluted with water to give a wash containing about 15% w/v of sugar, and to a winemaker this may come as a surprise, particularly when he remembers that the aim is the production of alcohol. It is not a difficult matter by 'feeding' the sugar to the must to finish with a wine of some 16% or more of alcohol v/v. The distiller, however, finds it more economical for the yeast to work at the height of its activity for two or three days and to produce about 7·5% v/v of alcohol than to let is slow down as the increase of alcohol inhibits its activity and to take perhaps a further two months in order to double the amount of alcohol produced from a stronger sugar solution.

Time and space are money where the demand and supply are for thousands of gallons of alcohol, and the distiller limits his choice between a faster fermentation, say two days, of a weaker mash that thereby calls for more and larger containers, and a slower fermentation, say three days, of a stronger mash, whose bulk takes less space but demands more time for working out. In theory, 15% w/v of glucose should give about half its weight, or a little more, in alcohol, viz. 7·6%, in accordance with the molecular weights of the substances involved:

$$C_6H_{12}O_6 \rightarrow 2C_2H_5OH + 2CO_2$$

molecular weights 180 $\qquad 2\times46=92 + 2\times44=88$

In practice, for reasons discussed elsewhere, it is rather less than the amount expected in theory, say 90% of this, = 6·9%.

Beet molasses contain most yeast nutrients, but in the case of cane molasses it is usual to add a little ammonium sulphate and ammonium phosphate, though not to the extent of encouraging unnecessary yeast growth at the expense of the sugar; these points are of little concern to the winemaker thinking in terms of a few gallons of wine, but they are vital to the producer who is competing with the synthetic alcohol of petroleum chemicals. The pH of the must is brought to about 4·5 by means of adding sulphuric acid, as this is a figure encouraging to the glycolytic enzymes and discouraging to unwanted bacteria. In America, where grain is the usual ingredient for industrial alcohol, the pH is adjusted by 'lactic souring', i.e. by the addition of lactic acid, or more commonly by means of a grain infusion culture of lactic bacteria, such as *Lactobacillus delbruckii.* It seems that this practice is less common nowadays.

Infection by bacteria always has to be considered where fermentation is on such a gigantic scale, although this does not present any great problems with alcohol fermentation. Pasteurisation is an expensive process with such a bulk of wash, and the usual practice therefore is to boil a strong concentration of molasses and then dilute this with pure water after cooling off. Any bacteria that survive, such as the spore-producing species of *Clostridia*, are then controlled by the 4·5 pH of the mash, in the acidity of which they cannot reproduce. Acetobacter that might find access after pasteurisation are controlled by the layer of carbon dioxide above the medium. Thus heat, acid, carbon dioxide and the short period of fermentation are all aids in the suppression of infection.

A *strong-attenuation yeast* ('attentuation' is the fall in gravity caused by conversion of sugar to alcohol) is chosen, attention being paid not so much to its *high* alcohol tolerance, for the amount produced is not so very high, as to its ability to work with the least inhibition from the alcohol formed. Speed and good returns are the factors guiding the choice, and the yeast looked for will maintain these in stronger sugar solutions than its competitors. 100,000-gallon tanks of wash need very large quantities of yeast, and various systems exist for minimising the labour involved. Usually these are based on a series of 'starters' or increasing size, from the initial agar-slope in a test-tub to a final tank containing some hundreds of gallons of starter medium. Increasing attention is being given nowadays to the value of fresh cultures at short periods, because of variants and mutations that cause deterioration of species in subcultures. The 'continuous yeast process' as opposed to the older 'batch method' is also attracting interest.

3. Alcohol Distillation

The Theory

SEPARATION. The separation of two liquids that have been mixed together presents no difficulties if they are *immiscible*, that is, they are insoluble in each other. Thus, for instance, if oil and water are mixed they do not combine but remain in separate globules, so that all that is necessary to separate them in the laboratory is to allow the mixture to remain undisturbed in a 'separating funnel' until there is a clear demarcation line between the liquids, when the lower layer can be run off through the funnel's stem.

SIMPLE DISTILLATION. The separation and recovery of a liquid from a substance completely dissolved in it is also a simple matter provided that the substance is non-volatile, and the process is usually loosely called 'simple distillation'. For example, if a solution of salt in water is heated to boiling point the water alone vaporises, leaving the non-volatile salt behind. In the laboratory the solution is poured into a distilling flask closed with a cork containing a thermometer, the latter raised sufficiently to be in the vapour and out of the liquid. The vapour passes out of the flask via its side tube and through a condenser such as the 'Liebig condenser'. This consists of a glass tube surrounded by a glass water-jacket, with an inlet and outlet for a stream of cold water to flow continuously. As a result, the vapour is cooled and condenses as it passes through the tube, being collected in a 'receiving flask'. Distilled water, free from salts, is prepared in a similar way for use in our car batteries, and drinking water from sea-water for use when ships are at sea.

Difficulties begin to arise when the aim is to separate two volatile liquids that are miscible in the liquid phase. They then dissolve in each other, and remain as a single phase, without settling out on standing, as do oil and water. Further, both are now volatile and capable of vaporisation, so that a simple distillation as used with salt and water does not give the same satisfactory results.

The difficulties involved may not at first be apparent. The principle of separating two volatile liquids by distillation turns on the fact that both have different boiling points, and one will thus be more volatile than the other; thus if a mixture of water and alcohol is to be distilled we are dealing with a liquid mixture whose two components have a difference of 22° C between their boiling points. The b.p. of water is 100° C, whereas that of ethyl alcohol is considerably lower at 78·4° C, so that the latter vaporises more easily. Impurities in the water may raise its b.p. a trifle, but we can disregard this, as also the variations caused by differences in atmospheric pressure and the height of the barometer. Why, then, it might be asked, should not the temperature of the mixture be raised carefully to 78·4° C, and held there while the alcoholic vapour that distils over at this temperature is condensed via the Liebig condenser as a simple distillation, eventually leaving pure water in the distilling flask and pure alcohol as the distillate?

Unfortunately Nature has not arranged matters in such a facile way, and a few observations will show that distillation in such a situation is a much more complex process than at first appears. Let us assume that we are dealing with a mixture of A and B, two volatile miscible liquids whose *complete* separation is feasible by distillation. B represents the more volatile liquid, which vaporises at a lower temperature than does A.

1. The mixture will have a new boiling point, lying somewhere between the b.p.s of the two components. The larger the proportion of the liquid A, with its higher b.p., the higher will be this new b.p., and vice versa.

2. When the mixture is heated at its appropriate b.p. the vapour contains a proportion of *both* substances, even though this temperature is below that of the b.p. of A. Water, for example, boils at 100° C, but it vaporises at much lower temperatures, even room temperature; if this were not so, ground would never be dry after rain.

3. The composition of this vapour, however, will not be the same as that of the liquid phase from which it is distilling at any one temperature. Obviously, the more volatile B will vaporise more easily, so that the vapour will be richer in this component than the liquid mixture at the same temperature. It is this fact, indeed, that makes distillation a feasible means of separating the two liquids.

4. As B vaporises more easily than A, it follows that the liquid mixture becomes progressively poorer in B during the distillation. In other words, it contains a larger proportion of A as the process proceeds, and therefore its boiling point will rise towards that of A, until it finally reaches it, when the distilling flask will contain pure A.

5. As the liquid mixture thus increases in its percentage of A and its b.p. becomes higher, so more of A is given off in the vapour, and this, too, increases in its percentage of A, the latter predominating

in later fractions. Nevertheless, as long as any of B is left in the flask, the vapour still contains more of B in its composition than does the liquid mixture at that temperature.

The final result of such a simple distillation is that pure A is left in the distilling flask, but the distillate in the receiving flask is still a mixture of A and B, though more concentrated in B than originally —witness the amount of A that has been separated from it. If a complete separation of B is wanted it will be necessary to redistil the condensed mixture of A and B, called a 'fraction', several times, at lower b.p.s each time, so increasing the concentration in the distillate of the more voluble B. At best, this is a tedious and cumbersome process, and a means of linking up the distillation of these separate fractions is desirable; this is found in *continuous fractional distillation.* Before this is discussed, it will be of assistance to give a diagram of the distillation that has just been described.

EQUILIBRIUM DIAGRAMS. The graph over page, known to distillers as a *phase diagram* or *equilibrium diagram*, shows the liquid curve and the vapour curve of a mixture of two volatile miscible liquids; A with a b.p. of 140° C, and B with a b.p. of 100° C. The definition of the boiling point of any liquid is the temperature at which the liquid is 'in equilibrium' with the vapour at a pressure of 760 mm. That is to say, it is the temperature at which the liquid begins to boil and the vapour to condense, so that a state of equilibrium is reached between the liquid and vapour phases. The two curves on the graph show the composition of the liquid and the vapour for any temperature where they are in equilibrium in this sense.

From the graph, if our mixture consists of 20% B (and 80% A) its b.p. can be found by drawing a vertical line from the figure 20 on the Composition Axis until it cuts the liquid curve, then drawing it horizontally until it cuts the Temperature Axis at about 124° C, the required figure. Notice where the horizontal line cuts the vapour curve, and drop a perpendicular line to the Composition Axis; the vapour coming off at this temperature has a composition of 65% B (and 35% A). Hence we can see that at its b.p. of 124° C a liquid containing 20% by weight of B is in equilibrium with a vapour containing 65% of B. Similarly, the b.p. of any composition of A and B can be found, and the composition of the vapour given off at any temperature.

Assume that a small portion of pure B is wanted from our mixture of 20% B and 80% A. If the first fraction that distils over at 124° C is collected it contains, as shown by the graph, 65% B. This is then

Equilibrium diagram

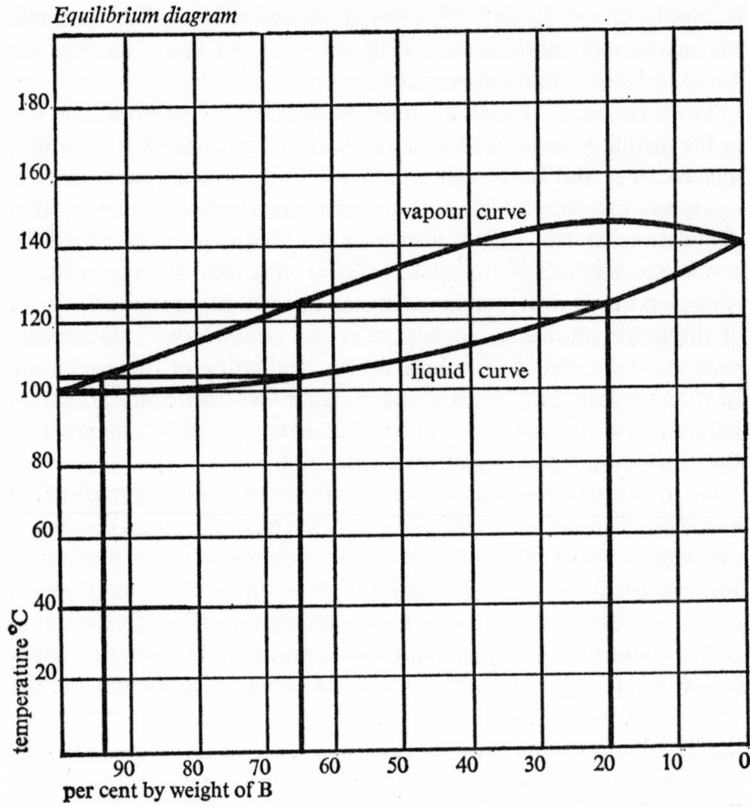

If the graph consists of a single loop, with liquid and vapour curves joining in this way on each side of the graph at the boiling points of the constituents, the liquids can be completely separated by distillation

redistilled at its new b.p. of about 107° C, a figure found from the graph as indicated above, to give a vapour now containing 95% B. This fraction is collected and redistilled, and so on until pure B is obtained. These tiresome repetitions are avoided in practice by 'fractional distillation'.

FRACTIONAL DISTILLATION. In order to link up the stages of simple distillation into a continuous process, a fractionating column is used, whereby the evaporation and condensation of the distilling flask and condenser are repeated continuously *within the column.* In the laboratory this may be a glass tube packed with glass balls, sections of glass tubing, cylinders of porous pottery or perhaps spirals of metal foil. It is inserted in the neck of the distilling flask, and from a side tube near the top of the column the vapour passes as before into the Liebig condenser. The thermometer is inserted in the stopper that closes the column.

As the vapour commences to leave the mixture it is richer in the volatile substance B than is the mixture in the flask, but it still contains considerable amounts of the substance A. The column is much lower in temperature at the top, where the condenser is removing heat, than at the bottom, where the heated flask is supplying it, and this variation of temperature from top to bottom is gradual. Consequently, as the vapour enters the column it is cooled slightly and some of it condenses. As A is less volatile than B and its boiling point is higher, it is more inclined to condense than B, so that the composition of the vapour as it ascends a little higher up the column is rather richer in B than it was when it left the flask. The same thing happens again and again as the vapour mixture continues up the even-cooler portions of the column, so that it becomes increasingly greater in B content and lower in A. In addition to this, the heat of the fresh vapour ascending tends to re-evaporate any of B that has previously condensed, and also the coolness of the liquid descending encourages more condensation of A.

By referring to the equilibrium diagram of A and B, the composition of the liquid trickling back and of the vapour ascending can be ascertained for any temperature at which they are in equilibrium. As the temperature progressively drops higher up the column, so it will be seen that the vapour is increasingly high in B. If the fractionation is carried out efficiently pure B should pass into the condenser. For this to take place the distillation should be conducted very slowly, and there should be efficient contact over as large an area as possible of the liquid and vapour phases, so that equilibrium may be successfully attained in every portion of the column.

AZEOTROPES. Distillation is further complicated by the fact that mixtures of miscible liquids fall into two main groups:

a. In the first, the new b.p. lies somewhere between that of the two components, according to the composition of the mixture. As the vapour always contains a greater proportion of the more volatile component than does the liquid phase with which it is in equilibrium, complete separation of the components is possible by means of a satisfactory fractional distillation apparatus. The distillation just described is an example of this type of mixture.

b. In the second group what are called 'maximum or minimum boiling-point mixtures' are formed. At such boiling points, when the composition of the mixture reaches a certain balance the vapour given off is of the same composition as the liquid, so that from this point of view the mixture then behaves *as though it were a single pure substance*, and it is called an *azeotrope*. Therefore it is impossible to separate the components further by fractional distillation, and the azeotrope thus bars the way to complete separation.

A mixture of alcohol and water is typical of systems that form a 'minimum boiling-point mixture', not an uncommon occurrence where one of the liquids contains the hydroxyl group (–OH), the distinctive group of alcohol. Ethyl alcohol forms a 'constant boiling mixture' with water, or azeotrope, when the composition is 95·6%

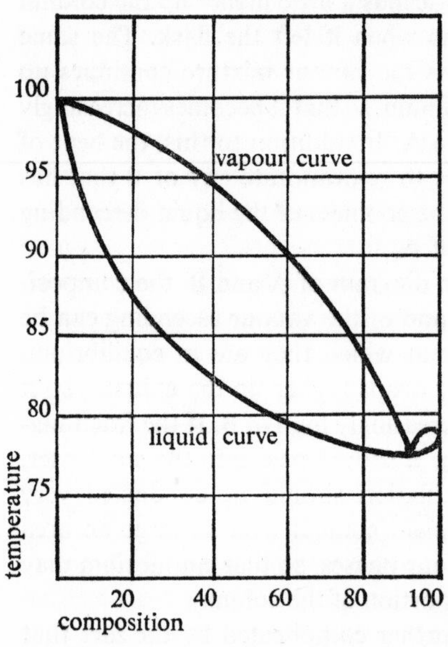

Equilibrium diagram for a minimum boiling-point mixture

When the vapour and liquid curves do not join at each side of the graph at the boiling points of the constituents, but meet to form a double loop, the point where the two loops meet represents the azeotrope; complete separation is not then possible by distillation

Azeotropic Composition of the System Water and Ethanol

At the minimum b.p. 78·2°C of EtOH/H_2O at 760 mm Hg pressure, the liquid contains 90 moles % EtOH (Noyes).

Mol wt H_2O = 18

Mol wt EtOH = 46

∴ 10 moles H_2O = 10 × 18 = 180 g H_2O

90 moles EtOH = 90 × 46 = 4140 g EtOH

∴ % H_2O w/w = $\frac{180 \times 100}{4140}$ = 4·4 %

∴ % EtOH w/w = 95·6%

by weight of alcohol and 4·4% by weight of water, boiling at the minimum boiling point of 78·2° C at normal pressure.

The result of fractional distillation of this mixture is that the vapour condensing from the head of the fractionating column contains this same composition, namely 4·4% of water, and it is impossible to reduce this amount by further distillation. If a series of simple distillations is carried out as an alternative to fractional distillation the b.p. of successive concentrations of the mixture will drop until 78·2° C is reached, and as this is the minimum boiling point possible, again the vapour will contain 4·4% of water in common with the liquid. Thus, pure alcohol is unobtainable by distillation, unless it is carried out at much lower pressures than atmospheric. A distillate containing 95·6% by weight of alcohol is known as 'rectified spirit'. An equilibrium diagram for a mixture of alcohol and water is on facing page, and a study of the vapour and liquid curves will clarify the points made.

Absolute alcohol, or 'anhydrous alcohol', free from any water, may be prepared from rectified spirit in either of two ways. The first method is to use quicklime as a dehydrating agent. Alcohol is heated with lumps of quicklime for six hours, or left corked for several days, until the water combines with the quicklime to form calcium hydroxide. It is then redistilled. About 0·2% water still remains, and this can be removed, if complete dehydration is necessary, by leaving it standing over metallic calcium, thus the decomposed:

$$Ca + 2H_2O \rightarrow Ca(OH)_2 + H_2$$

Calcium *Water* *Calcium hydroxide* *Hydrogen*

The other method is to distil the alcohol with an *entrainer*, such as benzene, b.p. 80·5° C. This process is based on the fact that benzene forms a ternary azeotrope with alcohol and water boiling at 69·7° C, and also a binary azeotrope with alcohol boiling at 72·5° C. These can then be separated from the anhydrous alcohol required boiling at 78·4° C.

Commercial distillation is based on the theory that has been discussed, although the latter has been set out as experiments in the laboratory. It should not be imagined from this that the distillation of ethyl alcohol is a regular practice in college work. Some authorities quote '*de minimis non curat lex*' ('the law does not bother itself with trifles'), but the fact is that any distilling of alcohol without licence can never be regarded as a legal procedure, and educational grounds would not be accepted in law as a valid reason for so doing. It is not generally realised that every distillation apparatus, even the simple instruments in schools and those in pharmacies for producing distilled water, has to be registered with Customs and Excise, and from time to time the local officer pays a visit and inspects it carefully.

Commercial Distilling, The Practice

THE POT-STILL. The commercial equivalent of the simple distillation process is the pot-still. This method is used in the manufacture of whisky, brandy and rum, because of the need to carry over many of the by-products of the distilled liquid or 'wash'. The distinctive flavour of the first two spirits mentioned and that of certain rums comes, with some modification by the process, from the same source as the alcohol, in contradistinction to gin, which owes its individual flavour to external sources. The substances contributing to the distinctive flavours of these spirits are known as *congenerics*, and they are mainly esters, acids and higher alcohols; glycerine, a by-product of fermentation that has an important role in the composition of wine, does not distil over, for its boiling point is 290° C. In addition to ethyl alcohol, the wash may contain methyl alcohol and fusel oil. Methyl alcohol, from the pectin of the grape, has a boiling point of 66° C, so that it distils over in the first fraction, which is rejected by experienced brandy-distillers. Its presence in the raw and largely unmatured spirits made for home consumption by the French *bouilleurs de cru*, some of whom have the legal privilege to distil brandy and marc, is doubtless one of the reasons for the damage done to body and mind. Amyl alcohol, the main constituent of fusel oil, derived from the protein of the ingredients, has a boiling point of 131° C, so that it comes over at the end of the distillation, part of it remaining in the still with the waste liquor. Although unpleasant in itself and poisonous in large amounts, fusel oil in quite minute amounts contributes by means of esters, formed during maturation, to the flavour of brandy, whisky and some rums. Jamaican rum, for instance, which preserves a very high flavour from its original wash, has altogether only 8 parts per thousand of all three congenerics.

Basically, the pot-still is little more than a large kettle. Various improvements have been added in France, in the more commercialised Cognac district, although 150 miles to the south-east the small producers of Armagnac, the 'brandy of Gascony', keep the simpler stills or even wait for the *alambic ambulant*, the travelling still, to visit their farms. The first improvement is the use of a *chauffe-vin* a chamber that surrounds the tube through which the hot vapours pass from the still to the condenser. As the name implies, its purpose is to warm the cold wine, which is led through the chamber on its way to the still. The second variation is the construction of a large bulb in the pipe leading away from the still, just above the exit from the pot, so that this forms a kind of primary condenser and returns the less-volatile vapours back to the still. A third modification may be the division of the still into two parts, so that the vapours form the lower section bubble through the liquid in the upper chamber, with the result that the vapours leaving the latter are more volatile than would otherwise have been the case; it thus forms a simple sort of two-fraction column.

During the first distilling the first and last fractions, the 'heads' and 'tails', are removed, and the main fractions, the *brouillis*, are collected, leaving the waste, the *vinasse*, almost exhausted of alcohol. A second distilling follows, consisting of three such collections of *brouillis*, plus the 'tails' and a little 'heads' of an earlier secondary distilling. The main fraction of this, containing about 70% alcohol, will in the course of time constitute French brandy.

The malt-whisky pot-still is larger, but of the very simplest type without modification, and they exist in pairs, with a condenser linking them. The wash enters the larger of the two, called the 'wash-still', where it is heated over a peat and coke fire. The uneven heating of this direct firing causes sticking and 'bumping', so that an automatic stirrer, called 'a rummager', and consisting of four arms dragging chains, is used to rotate the wash.

At the end of the first distilling the strength of the liquid has about doubled, from 10 to 20% alcohol. The distillate, cooled in worm condensers, passes on as 'low wines' to the 'spirit-still' for the second distilling. As with brandy, the first fractions, called the 'foreshots', with their methyl alcohol, are removed; likewise the last fractions, called the 'after-shots' or 'feints', containing fusel oil. Both are returned to the spirit-still with the following distillate of low wines from the wash-still. As the second distillate comes over from the condenser, it passes through the 'spirit safe', a glass-windowed box that has been sealed by the Excise authorities, and it is by the visual inspection of this that the skilled and vital decision is made at what point to separate the main whisky fraction from the foreshots and feints. The pot-still malt whisky is matured at about 116° proof, say about 65% alcohol. Pot-stills are not used for whisky in America, patent stills being universal, so that their whisky is light-bodied and slight in taste and aroma.

THE PATENT STILL. The alternative name for the apparatus is the Coffey Still, commemorating Aeneas Coffey, the Irishman who patented it in 1831, although as with many discoveries it was being used about the same time elsewhere, in France. From the point of view of separating alcohol from the wash, it is a much more efficient apparatus, with a high power of attenuation, complete control over selection of fractions, economic operation of heating and, particularly important, a continuous process of distillation without the chore of recharging the stills when the distillate has been collected. In the name of efficiency, however, the still goes too far for the producers of brandy and malt whisky, for the spirit is rectified so completely that congenerics are removed with the water and stay behind in the wash, although it is true that adjustments can be made to the rectifier to admit a percentage of these if desired. Consequently, the manufacturers employing the patent still are those producing industrial alcohol, gin—a neutral spirit with flavourings or 'botanicals' added later, 'grain whisky'—the ambiguous name for the light-bodied spirit added to malt whisky in about 60% proportion to lighten the blend for Sassenachs, and those light rums that are flavoured later like gin.

The patent still consists of two columns, known as the 'analyser' and the 'rectifier'. Perforated plates divide the analyser into compartments, and the wash is introduced at the top so that it slowly percolates downwards past the plates. As it does so, it meets steam blown upwards from the bottom of the column, with the result that the volatile alcohol vaporises and is carried over by the steam into the rectifier. The spent wash, with some of the condensed steam, runs out of the bottom of the analyser.

The rectifier is really a development of the principle of the fractionating column, and in commercial practice it may consist of a packed column as in the laboratory, a plate column, or a bubble-hood or bell column. The last is commonly used in the rectification of alcohol. Trays are placed at intervals, containing 'bubble-caps' and 'weirs'. The latter ensure that the wash reaches a certain depth on the tray before passing on via the 'downcomer', and thus the ascending vapour bubbles through a layer of liquid, so ensuring good contact of vapour/liquid phases.

The action is essentially as set out in the previous section on distillation, with the spirit vapours and steam introduced at the bottom of the rectifier. The more volatile vapours ascend and the less volatile condense and return, so that different fractions form at different levels. One very important difference from the laboratory column is that the pipe bringing the cold wash to the analyser zigzags its way down the rectifier, thus at the same time cooling the vapours and being itself warmed in the process. Alcohol is taken off about three-quarters of the way up the rectifier, and the fractions condensing at the top and bottom are returned to the still.

Rectified alcohol, 95·6%, is produced, and if pure alcohol, as for internal-combustion engines, is wanted, an azeotropic distillation follows, probably with benzene. By-products, such as fusel oil, used as a solvent for cellulose lacquers, carbon dioxide and glycerine, are all carefully preserved. In this way, particularly when industrial alcohol is being produced, the fermentation process is able to compete successfully with the synthetic methods used in the petroleum industry. Improved methods of fermenting the liquor are also now well established, such as Benoit's 'continuous yeast process', where the yeast is recovered by centrifugal separators and re-used, resulting in a saving of some 5% in sugar that would otherwise be diverted to yeast generation by the old batch methods.

B. Synthetic Processes

It is surprising how few people realise that industrial alcohol can be manufactured other than by the process of fermentation, and even those who are aware of this alternative method of production seldom appreciate the vast extent of the industry. Until the end of the First World War alcohol was always obtained by the distillation of fermented solutions, yet in 1956 a third of Britain's requirements of alcohol were supplied synthetically, and at present about two-thirds is made in this way in this country. The figures for 1960 for synthetic ethyl alcohol were 100,000 long tons. In America the percentage is 90, and it looks as if this will rise as the demands of industry increase. This does not mean that fermentation sources of supply are being ousted; the demand for industrial alcohol is such that the fermentation process continues to operate to full capacity.

What favours in particular the synthetic production of industrial alcohol is the uncertainty of the supply of fermentable material, such as molasses, grain and vegetables. If there is a surplus of these and transport costs remain low, then an efficient fermentation process becomes a keen competitor, but one cannot predict over a number of years what supplies will remain steady and what prices will be charged for the raw material. Consequently, the economics of the synthetic method are in its favour unless the raw materials are easily available in the country conducting the fermentation, and this is not by any means the case in this country.

Just after the First World War the British Government investigated the possibilities of utilising vegetable products grown in these islands as materials for producing alcohol in quantity to supplement petroleum supplies. As a result, the decision made was that it was not economically possible for several reasons: (1) insufficient acreage; (2) high cost of cultivation; (3) high cost of manufacture; (4) raw materials most suitable were also important foodstuffs. Of the three vegetables considered as potential suppliers of starch, potatoes, mangolds and Jerusalem artichokes were approved as suitable, and oddly enough the last named was considered to be the best for this purpose. (It contains *inulin*, a polyfructoside.)

1. Alcohol from Ethylene

For the demands of industry and power for alcohol, the synthetic methods based on the conversion of ethylene have proved a practical proposition. Ethylene is a colourless gas obtained from two main sources: natural gas, or petroleum refinery gas, and liquid petroleum. In the United Kingdom no natural gas has been discovered in the quantities needed, and the refinery gas from the distillation of crude petroleum is being increasingly used for preparing special fuels and additives. Thus by far the greater part of ethylene originates as a by-product from the cracking plants and units, where the 'reduced crude' is subjected to thermal cracking or *pyrolysis*, a term used to denote the rupture of large hydrocarbon molecules into smaller ones.

The normal synthesis of ethyl alcohol from ethylene is the *sulphuric process*. The ethylene is absorbed into 90–98% sulphuric acid at a pressure of 12–30 atmospheres and at a temperature of 85° C. As a result, the solution contains a mixture of monoethyl and diethyl sulphates. A hydrolyser then decomposes these mixed esters by interaction with water into ethyl alcohol, dilute sulphuric acid and small quantities of ethyl ether. Finally, a stripping column causes steam to remove the volatile components, leaving sulphuric acid at the bottom. Processes follow to purify and concentrate the crude alcohol vapour.

stage 1

C_2H_4 + H_2SO_4 → $C_2H_5OSO_2OH$
ethylene *sulphuric acid* *monoethyl sulphate*

$2C_2H_4$ + H_2SO_4 → $C_2H_5OSO_2OC_2H_5$
ethylene *sulphuric acid* *diethyl sulphate*

stage 2

$C_2H_5OSO_2OH$ + $C_2H_5OSO_2OC_2H_5$ + $3H_2O$ → $3C_2H_5OH$ + $2H_2SO_4$
monoethyl + *diethyl sulphates* *water* *ethyl alcohol* *sulphuric acid*

The *direct hydration process* is a more recent method of synthesising ethyl alcohol from ethylene. The ethylene is compressed to 68 atmospheres and mixed with water. In the reactor the mixture is passed in the vapour phase at 300° C over a catalyst of phosphoric acid on silica. Recycling of the ethylene is necessary, as only about 5% conversion is obtained with each passage through the reactor. Condensation and separation processes follow, and the alcohol is finally concentrated at 95·6%.

C_2H_4 + H_2O ⇆ C_2H_5OH
ethylene *water* *ethyl alcohol*

2. Alcohol from Acetylene

Ethyl alcohol can be produced from acetylene, a gas that can be obtained from calcium carbide by the addition of water. Therefore coal is the indirect origin of acetylene, since the manufacture of calcium carbide is based on heating coke and quicklime together in a closed electric furnace to a very high temperature of over 2000° C. An alternative source of acetylene is from hydrocarbons where these exist, largely natural gas consisting essentially of methane.

The route of synthetic alcohol from acetylene is via acetaldehyde. Catalytic hydration of the acetylene, using sulphuric acid in the presence of mercuric sulphate, produces first vinyl alcohol, which then rearranges itself to give acetaldehyde. This incorporation of water—'hydration'—is at atmospheric pressure at 95° C. The addition of hydrogen by catalytic hydrogenation then converts the acetaldehyde into ethyl alcohol.

stage 1

C_2H_2 + H_2O → $H_2C{=}CHOH$ → CH_3CHO
acetylene *water* *vinyl alcohol* *acetaldehyde*

stage 2

CH_3CHO + H_2 ⟶ C_2H_5OH
acetaldehyde *hydrogen* *ethyl alcohol*

Chapter 12 The Measurement of Alcohol Content

To ascertain the alcohol content of wine, there are several methods at the disposal of the chemist.

A. The Ebullioscope

The purity of a substance can be ascertained by measuring its melting or boiling point and comparing this with the known normal figures. Impurities present, even in minute amounts, affect the b.p. or m.p. of the substance, usually raising the first and lowering the second.

By means of an apparatus called the ebullioscope, the chemist can in this way discover the alcohol content in a mixture of alcohol and water by the regular variation in the boiling point as related to the proportions of the two constituents in the mixture. By the use of a sliding scale, that is part of the outfit, he converts his reading to percentage of alcohol present.

B. Chemical Analysis

Quantitive analysis is another method of ascertaining the alcohol content. This involves the use of a chemical to produce a reaction with the ethyl alcohol present, and dichromate is often used for this purpose, because in an acid solution it oxidises alcohol to acetic acid.

C. The Hydrometer

This might seem to be a method available to the amateur, but as has been explained, it is of no use for the direct determination of the alcohol content of a bottle of wine, because we are dealing here with a complex solution containing a number of substances. The employment by the chemist of the hydrometer to this end involves the process of distillation in order to separate the liquid from the sugar and salts; further, it is usual to decompose volatile esters, which might pass over with the alcohol and disturb the precise measurements, by heating with an alkali prior to distillation. Clearly, these activities call for training and cannot be lightly undertaken by the average winemaker. The only practical means of ascertaining alcohol content is by measuring the amount of sugar converted during the fermentation, and this involves a knowledge of the proportion of sugar in the must at the start.

When this information is not available there is a direct way for the amateur winemaker to test the alcohol content of a wine by means of a little instrument called a 'vinometer'. This consists of a short glass tube of very narrow, hair-like bore, with a small funnel at the top. A few drops of wine are inserted in the funnel and allowed to run through the tube. This is then inverted and the fluid column falls, a reading in percentage of alcohol being read off the scale on the tube when the fluid stops falling.

The principle on which the vinometer is based is known as *capillarity*. The molecules of a liquid have a number of different kinds of forces producing cohesion, but the force that concerns us here is the force of attraction due to the motion of the electrons, and extending over minute atomic distances. By means of this, a molecule in the centre of a liquid receives a pull in all directions from other molecules at the same time, so that the uniform distribution of this cohesional power results in a zero force for the molecule in question. Where, however, the molecule is on the surface of a liquid, the cohesional position is different, because it is not surrounded by molecules in such a position. Here the force is not zero but downwards, because although the side pulls are balanced, there is no upward pull to balance the strong downward pull.

Thus on the surface of a liquid there is, as it were, a skin or layer, and the forces concerned in this are described as the *surface forces* producing the phenomenon of *surface tension*. The surface molecules have thus a higher potential energy than those within the liquid, and anything perpendicular to the surface, such as the sides of the vessel containing it, or a very thin tube touching the surface, attracts the surface molecules of the liquid and causes them to rise up it. As a result of the first situation, we have the typical curled-up shape where water meets glass, known as the *meniscus*, and of the second, the rise of water up a fine bore, or *capillary*, tube. Ethyl alcohol lowers the surface tension of water, and advantage is taken of this fact in estimating alcohol content of wine.

At its best, a vinometer is only approximate, and performs best with dry wine. Sugar in solution interferes with its readings to such an extent that it is really of little value in testing a sweet wine.

Proof Spirit

The scale for measuring the strength of spirits in England and America is peculiar in being based on a unit called 'proof spirit'. The term is said to originate from a rough-and-ready test before hydrometers came into general use. It was considered that potable spirits should contain about 50% pure alcohol, and proof that this standard had been attained was made by damping gunpowder with the spirits under test. If the gunpowder still ignited the alcohol content was satisfactory; if not, the spirits were 'under proof' and contained excess water.

The increasing importance to the Government of revenue from the duty on spirits called for more exact standards of measurement, and the Spirits Act of 1816 adopted Sikes's hydrometer for official purposes and gave for the first time an official definition of proof spirit—'that which weighs exactly 12/13ths of an equal measure of distilled water'. The temperature, an important element in such a definition, was not given in the original Act, but it is considered that the intention was that it should be 51° F, and the Customs and Excise Act of 1952 states, 'Spirits shall be deemed to be proof if the volume of the ethyl alcohol contained therein made up to the volume of the spirits with distilled water has a weight equal to that of 12/13ths of a volume of distilled water equal to the volume of spirits, the volume of each liquid being computed as at 51° F.' All of which boils down to the fact that proof spirit contains 48·24% alcohol by weight, or 57·06% by volume, at 51° F. Sometimes these proportions are quoted at 60° F, in which case they are 49·28% alcohol by weight, and 57·1% by volume; the S.G., as compared with water at this temperature, is 0·91976.

The scale, followed by the Sikes's hydrometer, is then divided up into 100 parts or degrees, with 0° for pure water, and 100° for proof spirit, just defined. Absolute alcohol, i.e. 100% alcohol, is thus represented by 175·1° proof. In speaking of the alcohol content in terms of this scale, we can express the same thing in one of three ways. A bottle of spirits sold in Great Britain must, be the law of January 1st, 1948, have its alcohol content expressed on the label, and this is shown as 70° *proof*, but it could also be written as 30° *under proof*, or as 70° *Sikes*. Since 100% pure alcohol is 175·1° proof, an easy conversion from one scale to the other can be obtained by remembering that the first is $\frac{100}{175}=\frac{4}{7}$ of the other:

Degrees proof to percentage by volume: multiply by $\frac{4}{7}$
Percentages by volume to degrees proof: multiply by $\frac{7}{4}$

America also uses proof spirit as a unit, but conversion from one scale to the other is very much simpler, since they take 100° proof spirit as containing 50% alcohol by volume at 60° F, instead of our awkward figure of 57·1%, although the alcohol in this definition is not quite anhydrous, having a S.G. of 0·7946 as compared with 0·79359 for absolute alcohol. Since American spirits are sold at various strengths commencing at 86° (U.S.A.), Scotch Whisky for export is bottled at 75° (G.B.)=85·6° (U.S.A.) on the American label:

$$\frac{85{\cdot}6^\circ \text{ U.S.A.}}{2} = \underset{\text{alcohol}}{42{\cdot}8\%} \times \frac{7}{4} = 74{\cdot}9^\circ \text{ G.B.}$$

France is more concerned with wine than spirits, as are most European countries, and consequently they do not use proof spirit as a unit, but employ percentage by volume as the standard of measurement, so that the scale ranges between 0% for water and 100% for absolute alcohol. An alternative means of expression is the scale of the Gay-Lussac hydrometer, named after its inventor, a French physicist. The only difference between the two is that the latter divisions are in *degrees*, so that French brandy containing 40% alcohol may have 40° *Gay-Lussac* in the catalogue.

On the whole, the other European countries follow France, although there are slight differences. Italy and Russia calculate the percentage by volume at a different temperature. Germany calculates percentage of alcohol by weight. Holland's proof spirit contains 50% alcohol at 59° F. It does seem unfortunate that conformity between the nations on this matter has never been achieved.

Section 4 Acids

Chapter 13 The Acids of Wine

The acid content of the must plays a most important part during the course of fermentation as well as during the later maturation of the wine. In the first instance bacterial infection is checked to an increased extent, and in the second esterification is encouraged; a correctly balanced acid content is, of course, fundamental to the flavour of a quality wine.

The acids derived from fruit juices used in winemaking are mainly citric, malic and tartaric acids, and to a lesser degree oxalic acid. Those formed as side-products of fermentation are acetic and lactic acids. Each of these will now be discussed in turn.

A. Acids from Fruits

1. Citric Acid

This is the acid particularly associated with the citrous fruits, hence its name, where it constitutes more than 95% of the total acidity, but it also predominates in such fruits as currants, pineapple and tomatoes. Currants are high in citric acid content, and V. S. L. Charley in *Chemical constituents of Fresh Juices*, 1936, gives 3·40% acidity by weight for Baldwin black currants, and 3·20% for the Boskoop variety. Red currants have rather less acid, as have white. Malic acid is also contained by currants, with a trace of oxalic acid, the proportion of these three acids being about 32 : 4 : ½ w/w. The figures for Hawaiian pineapple given as anhydrous citric acid, are 0·6%, while those for tomatoes are 0·2–0·6%, depending upon such factors as variety, climate and maturity. Even in citrus fruits malic acid also occurs, with small amounts of oxalic acid in the peel and edible portions, and the same situation will be seen in most fruit juices, namely either citric or malic predominating, together with an appreciable quantity of the second acid, and traces of others.

First isolated from lemon juice in 1784, citric acid production until about forty years ago was centred in Italy, particularly Sicily, where it was obtained from lemons unsuitable for export. After neutralising the juice with calcium carbonate, the resultant calcium citrate was decomposed by adding sulphuric acid, filtered and then evaporated to give crystals of citric acid.

In 1923 an American factory started producing citric acid from carbohydrates by using a mould and this has become the standard method of commercial production. This microbiological reaction may be written as

$$3C_6H_{12}O_6 + 9O_2 \longrightarrow 2C_6H_8O_7 + 6CO_2 + 10H_2O$$

glucose *oxygen* *citric acid* *carbon dioxide* *water*

but the equation is misleading, for it conceals a most complicated series of changes. The black mould *Aspergillus niger* is now the standard mould culture used in citric acid fermentation. In order to produce the required acid in quantity, a deviation from the Krebs cycle, as it is called, is necessary, and the normal sequence of enzymic activity has to be diverted from its main end. A similar instance is the blocking of the sugar-to-alcohol route in order to produce glycerine as a side-reaction instead of the usual end-product of alcohol. The deviation is started basically by one of two methods: either by restricting severely mineral elements, particularly of iron, manganese and copper, or by limiting the supply of phosphates. The carbohydrate used commercially is usually beet or cane molasses, the preference being for the former as being less contaminated with iron. If the latter is used it is likely to be the purer 'High Test' molasses rather than blackstraps.

The recovery of the citric acid from the medium is similar to the process used for lemon juice. The wash is neutralised with lime, after the fungus mat has been separated by filtering, and the resultant calcium citrate is decomposed with the aid of sulphuric acid into citric acid and calcium sulphate. The latter is precipitated, and the citric acid in solution is then evaporated to form crystals.

2. Malic Acid

The name of this acid derives from its occurrence in apple juice (Latin *malum* = apple), where it is the main acid present, although there is a small amount of citric acid and traces of others, such as quinic acid—particularly in cider apples—and lactic acid. The quantities vary according to variety, growing conditions and location, but the following figures are for some of the well-known varieties of English-grown apples, adapted from V. S. L. Charley's and T. H. J. Harrison's *Fruit Juices and Related Products*:

	Variety	*Malic acid,* %
Culinary	Bramley's Seedling	0·9–1·3
	Lane's Prince Albert	0·87
	Newton Wonder	0·55
	Cox's Orange Pippin	0·62

	Variety	Malic acid, %
Dessert	Laxton's Superb	0·53
	Blenheim Orange	0·53
	Worcester Permain	0·27
Cider	Kingston Black	0·68
	Frederick	1·00

In comparison with these, the acidity of pear juice, about 0·2% is remarkably low.

Malic acid is also the principal acid present in cherries, together with a small amount of citric and lactic acids, the figure for Morello cherries being 1·86% total acidity; this is three or four times the quantity present in the sweet varieties, which are low in acid. The following figures for acidity expressed as malic acid in English berry-fruits are from V. S. L. Charley's report on fruit juices at the Long Ashton Research Station in 1932, *Investigations in Fruit Products*:

		%
Green Gooseberry	Keepsake	1·96
Red Gooseberry	Ironmonger	0·80
Strawberry	Royal Sovereign	0·95
Raspberry	Baumforth A.	1·57–2·23
Blackberry	Mixed Seedling	1·42
Loganberry	—	2·70

Malic acid shares with tartaric acid the position of being the main acid of grape juice, supported by small amounts of citric, succinic and lactic acids. According to Amerine and Joslyn, the composition for Californian grapes is tartaric acid 0·2–0·8%, and malic acid 0·1–0·5%. The amount and proportion of the two principal acids vary considerably according to the variety of grape, the geographical latitude and the conditions prevailing during the season of ripening. Malic acid predominates at the commencement of the season, but the total acidity falls during the period of ripening, and this is primarily due to a decrease in the malic acid content; it forms an important factor in cell nourishment, and as the grape cells mature so the malic acid is consumed by them. Consequently, where grapes are grown in warmer climates, the result will be that tartaric acid is the predominant acid, although it should be noticed that ripening also decreases the amount of free tartaric acid and increases the amount found in the form of its salt, potassium bitartrate. In cooler climates, however, such as those prevailing in German grape-growing districts, the malic acid may remain the predominant acid, the resultant wine having therefore a marked acidity in its taste.

Malic acid content may also decrease by a quite different method, and at a quite different time. This is during the period of maturation after the fermentation has been completed, and in this case the malic acid is not consumed but is converted into another acid:

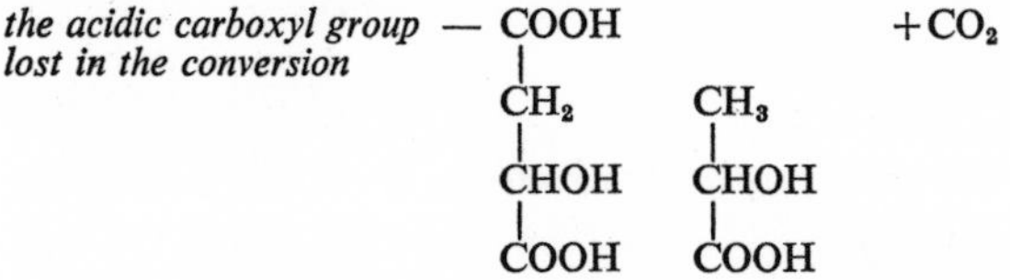

the acidic carboxyl group lost in the conversion —

$$\begin{array}{ccc} \mathrm{COOH} & & +\mathrm{CO_2} \\ | & & \\ \mathrm{CH_2} & \mathrm{CH_3} & \\ | & | & \\ \mathrm{CHOH} & \mathrm{CHOH} & \\ | & | & \\ \mathrm{COOH} & \mathrm{COOH} & \\ \textit{malic acid} & \textit{lactic acid} & \textit{carbon dioxide} \end{array}$$

The characteristic feature of these two acids is the group -COOH, called a carboxyl group, and consequently malic acid is 'dibasic', with two carboxyls, and lactic acid 'monobasic', with one carboxyl. To the palate, wine that contains malic acid has a more pronounced acidic effect than one containing the equivalent amount of lactic acid. In the well-known malo-lactic fermentation that can take place during maturation, Lactobacilli remove one of the two carboxyl groups of the dibasic malic acid, with resultant carbon dioxide being given off just as in the fermentation of sugar to alcohol that has preceded it. The equation above has been set out to make this change clear.

Such a reaction has no particular interest to grape wines of warmer climates, but where cooler latitudes have left considerable amounts of malic acid in the must, or where wines are being made from ingredients with high yields of malic acid, such as apples, this secondary fermentation is of considerable importance in the smoothing out of the matured wine. Sometimes some of the gas is purposely retained in the wine, so that it has the freshness associated with a slightly petillant wine. Further notes on this malo-lactic fermentation and the bacteria responsible for the reaction are given in the section on Wine Disorders.

3. Tartaric Acid

As it exists in nature, this is essentially the acid of grapes from sunny climates, although it is also found in juices of fruits such as mulberries and pineapples. As has been said, ripening decreases the amount of free tartaric acid and increases the quantity of potassium bitartrate, although it should be pointed out that the availability of potassium in the soil may also bear a close relation to this increase and account for variations in the percentage in grapes that are otherwise similar as regards their climatic situation.

Tartaric acid is like malic acid in being 'dibasic', with two carboxyl groups:

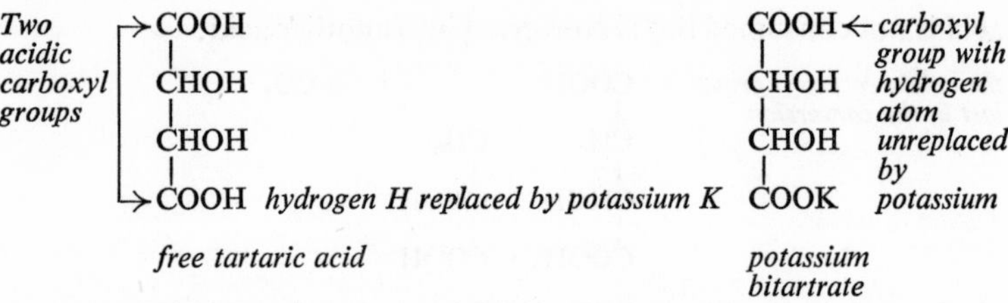

Unlike its related salt potassium tartrate, potassium *bi*tartrate results from only half of the acidic hydrogen atoms of the two carboxyl groups contained by tartaric acid having been replaced by potassium. (The *basicity* of an acid is the number of hydrogen atoms in one molecule of that acid which can be replaced by a metal: thus hydrochloric acid, HCl, is *monobasic*, sulphuric acid, H_2SO_4, is *dibasic*.) Consequently, it is also referred to as '*acid* potassium tartrate' or 'potassium *hydrogen* tartrate'. Its molecular formula is $KHC_4H_4O_6$.

It is only sparingly soluble, and consequently it crystallises out of grape wine that contains it to give what is known to vintners as *argol*. Low temperatures encourage this precipitation, and technical advantage is taken of the fact when excess tartaric acid is contained by the wine to store it in cool cellars or to make use of refrigeration processes. When purified, freed of colouring matter and recrystallised, it is known as *cream of tartar*.

Argol, which contains about 75% of potassium bitartrate, is the starting material for the commercial production of tartaric acid. It is boiled with dilute hydrochloric acid, and the solution treated with chalk and calcium chloride, with the result that calcium tartrate is precipitated:

$$2KHC_4H_4O_6 + CaCl_2 + CaCO_3 \rightarrow 2CaC_4H_4O_6 + 2KCl + CO_2 + H_2O$$

potassium bitartrate	*calcium chloride*	*chalk*	*calcium tartrate*	*potassium chloride*	*carbon dioxide*	*water*

The insoluble calcium tartrate is filtered off and decomposed with dilute sulphuric acid. This liberates tartaric acid, which passes into the solution and leaves calcium sulphate, which is only sparingly soluble:

$$CaC_4H_4O_6 + H_2SO_4 \longrightarrow C_4H_6O_6 + CaSO_4$$

calcium tartrate	*sulphuric acid*	*tartaric acid (in solution)*	*calcium sulphate (precipitated)*

Now the solution of tartaric acid can be filtered off and crystallised in vacuum pans. Further clarification with animal charcoal and purification follows, to give commercial tartaric acid in crystal form.

Summary of Terms

Cream of tartar:	purified potassium bitartrate
Tartar:	*crude* potassium bitartrate ('tartar' deposited on teeth is a salivary deposit of mucus and phosphate of lime)
Argol:	'wine stone' is grape wine lees containing some 75% of crude potassium bitartrate
Potassium bitartrate:	synonyms are: potassium hydrogen tartrate; acid potassium tartrate; bitartrate of potash.

4. Oxalic Acid

This is one of the oldest-known organic acids, and is stronger than most, although still weaker than the mineral acids. It is very sharp in flavour and extremely poisonous. Altogether, it is not an acid that is desirable in wine, and fortunately the only ingredient that is likely to bring it into the must in any quantity is rhubarb, still in common use for home-made wines.

The main acid of rhubarb juice obtained from the stalks is malic acid, ranging from 1·0 to 1·7%, according to the variety and the season when it is pulled, usually in the form of its acid potassium salt. Free oxalic acid is found in varieties of *Boletus* and, as its botanic name suggests, in the wood sorrel *Oxalis acetosella*, but more usually in nature in the form of potassium, ammonium and calcium salts.

The only safe parts of the rhubarb plant are the leaf stalks, and even these contain considerable quantities of oxalates. As used in making wine, after dilution with water, the juice from the stalks may be regarded as reasonably safe, and the writer has never heard of poisoning resulting from the drinking of this typical country wine. Nevertheless, it should be remembered that some people are allergic to oxalic acid salts, and any feeling of discomfort after drinking should be regarded as a danger in this direction. If the resultant wine is rough and over-acid precipitated chalk should be added to neutralise this acid, the wine then being carefully racked off from the insoluble calcium oxalate precipitated. The acid content should then be adjusted, if necessary, with the addition of a milder acid. Incidentally, chalk is the normal antidote to oxalic poisoning.

Leaf blades, of course, should on no account be used. These contain a higher percentage of oxalate, about 0·3%, and in cases where they have been boiled and eaten by unwary gardeners in place of spinach fatal poisonings have been recorded.

B. Acids from Fermentation

The formation of acetic acid and lactic acid as side-products of the Embden–Meyerhof fermentation cycle are covered in the chapter on the Biochemistry of Fermentation, so that there is no need to repeat here the reasons for their existence in wine. Abnormal quantities of both in wine can be caused by microbiological disorders, and these situations are discussed later in the section on Disorders.

1. Acetic Acid

In its pure form this acid is sufficiently powerful to burn the skin, although chemically it is regarded as rather a weak acid. It is not toxic, as is oxalic acid. Vinegar is a dilute solution containing about 7% acetic acid, although of course there should be bouquet and flavour associated with the ingredients from which it has been prepared.

Acetic acid can be prepared in a number of synthetic ways, as well as by fermentation.

A. From about the middle of the nineteenth century to the 1920s the commercial production of acetic acid was by means of the destructive distillation of wood. When heated in closed retorts at 300° F /80° C a liquid called *pyroligneous acid* is condensed from the fumes. As early as 1681 the German chemist Glauber had recognised its main content as 'acid of vinegar', and in 1661 Boyle found that 'wood spirit' was present, later to be classified as methyl alcohol. The process of wood-distilling has been covered in the section on alcohols.

B. Although this industry continues, assisted by the improvements in technique of the 'three-vessel system' that enables the separation in one operation of pyroligneous acid into tar, acetate liquor and wood spirit, the modern commercial production of acetic acid is from *acetaldehyde.*

This extremely important substance is made synthetically either from acetylene or from synthetic ethyl alcohol, a cheaper raw material Acetylene may be made from petroleum, as in the United States, Italy and Western Germany, or from coal via calcium carbide, as in this country. By means of passing acetylene into a solution of mercuric sulphate in hot dilute sulphuric acid its hydration takes place. First vinyl alcohol forms, which then rearranges itself at once to give acetaldehyde:

$$\underset{\text{acetylene}}{C_2H_2} + \underset{\text{water}}{H_2O} \rightarrow \underset{\text{vinyl alcohol}}{H_2C{=}CHOH} \rightarrow \underset{\text{acetaldehyde}}{CH_3CHO}$$

Synthetic ethyl alcohol, made from ethylene by the petroleum chemical industry as described in the section on alcohols, has its major use in the preparation of acetaldehyde. The latter is produced by the oxidation with air of the alcohol in the presence of a silver gauze catalyst:

$2C_2H_5OH + O_2 \longrightarrow 2CH_3CHO + 2H_2O$

ethyl alcohol *oxygen* *acetaldehyde* *water*

Whatever the origin of the acetaldehyde, its conversion to acetic acid is by further oxidation with either air or oxygen in the presence of metallic salts as a catalyst:

$2CH_3CHO + O_2 \rightarrow 2CH_3COOH$

acetaldehyde *oxygen* *acetic acid*

C. The process of deriving acetic acid from ethyl alcohol via acetaldehyde described above may also take place microbiologically with the aid of enzymic catalysts contained by certain bacteria. This is essentially what happens in the manufacture of vinegar, which is a dilute solution of acetic acid. The bacteria are *Acetobacter* of which twenty species exist. Although they are aerobes, that is they work with the aid of atmospheric oxygen, most of them can produce only partial oxidation of the alcohol, stopping when acetic acid is formed. In aerobic oxidation it is normal for this change to be complete, and indeed some species do continue the process to produce water and carbon dioxide. Strictly, fermentation is an anaerobic partial oxidation in which a balancing reduced end-product accumulates; the oxidation of ethyl alcohol to acetic acid obviously does not conform to this definition. Nevertheless, it is customary to apply the term 'fermentation' to this process, probably because in the first instance the expression included within its meaning the preliminary fermentation by yeast of sugar to alcohol.

The vinegar vat is one of the few places where the *Acetobacter* gets a welcome. The material to be fermented is normally malt wort in England, wine in France and cider in America, but whatever the basic alcoholic ingredient and whatever the technique employed, the fermenting agent is one or more species of *Acetobacter*.

In the manufacture of British malt vinegar the usual material is ground barley malt, though other cheaper grain is added for economy, such as flaked maize. The constituent starch is then saccharified, as with beer brewing, by mashing to activate the diastase complex of enzymes. The temperature in the mash tun is lower than with beer, for the vinegar manufacturer wants the minimum of unconverted dextrin left, and the wort is not boiled in the copper for the same reason. Maximum alcohol yield is the only end in view, and the relevant amylolytic enzymes are left unharmed to continue the final breakdown of starch to maltose even in the fermentation vat.

Without boiling and without hops, both aids of the brewer, to inhibit *Lactobacilli*, these tend to infect the wort, but on the other hand, the alcohol content is higher—about 6–7% w/v—always a safeguard against infection. The acid content will also eventually be higher than in a brewery, but not at this stage, and the result is that some maltose is diverted to lactic acid. No one appears perturbed about this, it seems. As the layer of carbon dioxide over the wort thins out, which previously helped to control bacterial growth by restricting the oxygen available, some *Acetobacter* will already have gained an early entry.

The bulk of acetification takes place in an *acetifier*, a wooden vessel holding perhaps 10,000 gallons, and packed with bundles of birch twigs on perforated floors. The wort trickles through slowly, with a maximum surface exposure to the air that enters through holes in the sides, so that oxidation by the *Acetobacter*, that collect on the twigs, is possible. The liquid is recirculated until the acid content is sufficient, and then withdrawn.

Probably a number of *Acetobacter* spp. take part in the acetification, the use of pure cultures being unusual in malt-vinegar manufacture. Two species are not welcomed. The first is *Acetobacter rancens*, which 'over-oxidises' alcohol to water and carbon dioxide, and the second is *Acetobacter xylinum*, that forms a capsule containing cellulose similar to that of cotton, and if not controlled forms the so-called 'blacksmith's apron' that clogs the twigs and blocks pipes with its growth.

In addition to malt, wine and cider vinegars, there are other forms such as 'spirit vinegar'. This is made by the oxidation of diluted alcohol, and as *Acetobacter* need nitrogenous food such as yeasts, malt extracts or autolysed yeast is added. Pure cultures are used, such as *Acetobacter curvum*. This is the strongest vinegar available, yet cheap to produce. 'Distilled vinegar' is malt vinegar that has been distilled and it should retain the original malt aroma.

2. Lactic Acid

The Swedish chemist Scheele, who also first prepared citric acid from lemons as calcium citrate, discovered lactic acid in 1790. It is formed by the microbiological fermentation of sugar, a process first noted by Blondeau in 1847, and Pasteur's studies of the organism twenty years later paved the way to commercial production, which started in 1881.

When milk turns sour it is the result of bacteria called *Lactobacilli*, which have converted the lactose, or milk sugar, into lactic acid. Sometimes putrefactive bacteria get an early start in the milk before there is sufficient lactic acid formed to inhibit them, and they break down the proteins, with the result that the milk 'turns bad', smells offensive and is dangerous to drink. Dilute lactic acid, on the other hand, is harmless, and claims are made for its beneficial qualities in the human diet. Yoghourt is made by inoculating milk with a species of *Lactobacilli* named *Lactobacillus bulgaricus*, the specific name deriving from Bulgarian shepherds who are credited with its original production. The mild acid content is claimed to restrict the multiplication of putrefactive bacteria in the intestines, where these are responsible for causing excessive flatulence and discomfort.

The conversion of the carbohydrate to lactic acid may be represented by the equation:

$$\underset{\textit{lactose or glucose}}{C_6H_{12}O_6} \longrightarrow \underset{\textit{lactic acid}}{2CH_3 \cdot CHOH \cdot COOH}$$

This is an over-simplification that omits the many intervening steps, for the fermentation follows the alcoholic fermentation pathway down to the production of pyruvic acid. Then, as the species of bacteria used commercially, *Lactobacilli delbrueckii*, contains no carboxylase enzyme, it is unable to split off the carbon dioxide, and the pyruvic acid is reduced to lactic acid:

$$\text{carbohydrate} \rightarrow \underset{\textit{pyruvic acid}}{CH_3 \cdot CO \cdot COOH} + \underset{\textit{hydrogen}}{H_2} \longrightarrow \underset{\textit{lactic acid}}{CH_3 \cdot CHOH \cdot COOH}$$

instead of giving acetaldehyde and alcohol:

$$\text{carbohydrate} \rightarrow \underset{\textit{pyruvic acid}}{CH_3 \cdot CO \cdot COOH} \xrightarrow{\textit{carboxilase}} \underset{\textit{acetaldehyde}}{\overset{\textit{reduced to alcohol}}{CH_3 \cdot CHO}} + \underset{\textit{carbon dioxide}}{CO_2}$$

Not all the species of *Lactobacilli* are restricted to lactose as the substrate, and the medium now used commercially for the production of lactic acid is usually beet molasses. This is a purer substance than raw cane molasses, since beet sugar is extracted by diffusion from the sliced beet, whereas sugar cane is crushed, so that cane molasses contains gum and cellular impurities. On the other hand, nitrogenous compounds are present in beet molasses that are difficult to separate later, and therefore the more expensive *High Test* cane molasses is an excellent alternative. This is the name given to a concentrated sugar-cane syrup that has been partially inverted. If used, it is necessary to add autolysed yeast, after the enzymes have been destroyed by boiling, to provide amino acids as a nitrogen source for the bacteria.

The diluted molasses is pasteurised at 90° C for fifteen minutes and reduced to the fermenting temperature of 48° C by further dilution. *Lactobacilli* are *thermophilic*, and need a higher temperature than most other fermentations. The optimum pH is 7·5, and this is adjusted accordingly. It seems curious that this acid-producing bacterium is not adjusted to an acid medium, and ground chalk to react with the acid as it is formed is therefore added, being continually agitated to prevent its settling to the bottom of the tanks. The fermentation is then started with the prepared culture, and in about 72 hours some 95% of the sugar will have been converted to lactic acid.

The fermentation is anaerobic, but there is no need with *Lactobacilli* to exclude air as with alcoholic fermentation, for its presence or absence makes no difference at all to the result. Infections are not such a danger here as with some fermentations, for the acid that forms is a protection, as is also the blood-heat temperature.

A solvent extraction of the lactic acid is now usual, superseding the older and more expensive crystallisation of the calcium lactate formed. After the wash has been filtered, dilute sulphuric acid is added to precipitate calcium sulphate, and the lactic acid in solution is then extracted with butanol as a solvent, filtered through charcoal and distilled.

Chapter 14 Sulphurous and Tannic Acids

A. Sulphurous Acid

Sulphur is an important element in life. Some of the amino acids contain sulphur, namely cysteine, cystine and methionine, and so do some of the vitamins, such as biotin and thiamine. Plants absorb it usually in the form of sulphates from the soil, and then synthesise the amino acids mentioned, which are in turn passed on to animals in their protein food.

As a disinfectant, the use of sulphur has been known in the wine industry for centuries, and there is an Act of Charles II forbidding its use in wine. A very common method of employment was the *méchage*, or sulphuring, of casks with sulphur candles. These are wicks impregnated with sulphur, which are lighted and lowered in metal containers into the cask. The burning sulphur combines with atmospheric oxygen to form the disinfectant sulphur dioxide, so destroying living micro-organisms in the vat:

$$S + O_2 \longrightarrow SO_2$$

sulphur   *oxygen*   *sulphur dioxide*

Sulphur dioxide dissolves easily in water to give sulphurous acid:

$$H_2O + SO_2 \leftrightharpoons H_2SO_3$$

water   *sulphur dioxide*   *sulphurous acid*

The double arrow in the equation indicates a reversible reaction; pure sulphurous acid cannot be isolated, since it decomposes into sulphur dioxide and water on concentration at normal temperatures. The reaction therefore never proceeds to completion, and an equilibrium is maintained.

Since sulphurous acid combines with oxygen to form sulphate and sulphuric acid, it possesses reducing properties, and is thus an antioxidant and a bleach:

$$2H_2SO_3 + O_2 \longrightarrow 2H_2SO_4$$

sulphurous acid   *oxygen*   *sulphuric acid*

Most bleaches oxidise the dyes to a colourless compound, but sulphurous acid bleaches by reducing the colour pigment. This effect is noticeable when adding sulphur dioxide to a coloured must, but later the colour returns by means of the action of atmospheric oxygen on the reduced pigment.

Sulphur dioxide may be employed in winemaking in three different forms: as a pressurised liquid, a solution or a salt. In commercial practice the liquid form of the gas is popular because of its easy handling. If the acid content of the must is low a dilute solution of sulphurous acid containing about 6% sulphur dioxide dissolved in water is an advantageous form of employment. The sulphite salts of alkali metals are available, since they are easily converted into sulphurous acid in the acidic wine must, although a small amount of acid content may be neutralised in the process. It should be borne in mind that a stock solution prepared either from sulphur dioxide and water or from a powdered salt and water decreases in strength when stored.

As sulphurous acid is a weak dibasic acid, it forms two series of salts: normal salts or sulphites, and acid salts or bisulphites. The normal sulphites are stable salts that liberate sulphur dioxide on treatment with acid; those of an alkaline metal, e.g. sodium sulphite (Na_2SO_3), are soluble in water and give an alkaline reaction. The bisulphites are either neutral or feebly acid in solution, and as solids exist as metabisulphites.

$$Na_2S_2O_5 + H_2O \longrightarrow 2NaHSO_3$$

sodium metabisulphite *water* *sodium bisulphite*

The sulphites are formed by the action of sulphur dioxide on oxides, hydroxides and carbonates, an excess of the gas producing the bisulphites. Thus, if a solution of sodium hydroxide is saturated with sulphur dioxide sodium bisulphite is produced in solution:

$$NaOH + SO_2 \longrightarrow NaHSO_3$$

sodium hydroxide *sulphur dioxide* *sodium bisulphite*

If the solution of sodium hydroxide is very concentrated and kept cold crystals of sodium metabisulphite may be precipitated:

$$2NaOH + 2SO_2 \longrightarrow Na_2S_2O_5 + H_2O$$

sodium hydroxide *sulphur dioxide* *sodium metabisulphite* *water*

The metabisulphite salts of potassium or sodium form a convenient way of introducing sulphur dioxide to the must for home winemakers; some make up a stock solution, others use Campden Tablets containing 0·44 g metabisulphite (7 grains).

Moulds and *Lactobacilli* are particularly susceptible to the effect of sulphur dioxide in the must, as are many wild yeasts. Its presence in the correct proportions does not produce a sterile medium entirely free of fungi, yeasts and bacteria, but assists in reducing the count of micro-organisms to an acceptable level during the dangerous period for infection preceding the formation of alcohol. Then as the antiseptic action of the sulphur dioxide decreases the strong fermentation of the yeast culture holds sway and alcohol is steadily formed, both factors in keeping down unwanted development of micro-organisms. The wine yeast, *S. cerevisiae* var. *ellipsoideus*, is resistant to the presence of sulphur dioxide, and thus allows a selective control to be exerted over organisms in the must. This resistance is probably due to the strong reductive powers of this yeast, a well-established phenomenon, and it has been suggested (Schanderl, 1952) that sulphur dioxide is reduced to hydrogen sulphide. Resistance is at its height when the yeast is in full activity, and the aldehydes formed in fermentation also assist in fixing the sulphur dioxide.

Not only is sulphur dioxide an effective disinfectant of vessels and a suppressor of unwanted micro-organisms but because of its reducing power mentioned above it can prevent or minimise oxidative changes in the must. Against this must be set the disadvantages that it has a distinct and unpleasant 'nose' and flavour. Laws in wine-producing countries are very strict in controlling the amounts to be used. France limits the proportion of SO_2 to 450 parts per million, Germany to 250 and Italy to 165. The claim is made that 300 parts is an amount easily detectable in wine as an after-flavour, but this is seldom reached by home winemakers, who normally add two Campden tablets to a gallon of must. This gives only 100 parts per million, quite a low proportion.

Sulphur dioxide has an affinity for aldehydic substances, and therefore a proportion in the must combines chemically with the glucose present to form bisulphite compounds, called hydroxysulphonates. This proportion is known as the 'combined fraction', leaving the rest, called the 'free fraction', as the only part which is active microbiologically against unwanted organisms. Another factor that affects the efficiency of the sulphur dioxide as a means of controlling fungi, bacteria and wild yeasts is the pH of the must. Sulphur dioxide in water solutions exists as the dissolved gas SO_2, sulphurous acid H_2SO_3, the bisulphite ion $^{-}HSO_3$ and the sulphite ion $^{--}SO_3$. Added to the must as a salt, metabisulphite is largely hydrolysed into sulphurous acid and bisulphite ions. The ratio of free sulphur dioxide to the combined is influenced by the pH as well as the sugar and aldehyde content. There is little free SO_2 at pH 7, it is mainly $^{--}SO_3$ and $^{-}HSO_3$, but as the pH falls the concentration of SO_2 rises. The amount of metabisulphite needed to be effective in a must with a pH of 3·5 can be reduced to a half or even a quarter of the quantity where the pH is 2·5 and still be satisfactory as a means of purification. Winemakers will therefore understand the inclusion of ½ oz of citric acid with six Campden tablets in a pint of water in recipes for a solution designed to sterilise apparatus.

In home-made wines the average amount of sulphur dioxide needed in a must is 100 p.p.m., but in juices low in acid this may need to be increased to 150–200 p.p.m. or, in terms of Campden tablets, 3–4 instead of the usual 2 per gallon of must. The free sulphur dioxide takes several hours before it reduces the micro-organisms to an acceptable level, and this satisfactory position is then held for another six to twelve hours, after which the perceptible control gradually diminishes. A suitable time to add the chosen yeast starter is between twelve and twenty-four hours after introduction of the metabisulphite; it is most important to bear in mind that sulphiting a must is by no means a permanent insurance against infection, but only a temporary measure that allows the desired fermentation to start under the best conditions.

Before leaving this subject it may be of interest to note the alternative *super-quattro* method used by some Italian vignerons. This utilises the alcohol itself as means of controlling wild yeasts in place of sulphur dioxide. The fermentation vat is filled two-thirds full with must and then has the remaining third topped up with completed wine of a 12% alcohol content. As a result, the must now contains 4% of alcohol, sufficient to inhibit the wild yeasts, but not enough to inactivate the wine yeasts. The weakness, of course, is that wine diseases flare up from time to time and contamination spreads through the winery.

B. Tannic Acid

Tannin is an essential ingredient of wine, its bitter astringent qualities conferring interest or 'zest' to the wine. The amount present distinguishes red from rosé and white wines, which have not absorbed so much tannin from the skin and stalks of the grape, and also varieties of red wine from one another. Claret, the red wine from Bordeaux, contains more tannin than Burgundian wines, a characteristic that leads partisans of the Côte d'Or to speak disparagingly of them—as might be expected—as having a '*goût de poussière*', a dusty taste.

Medicinally, advantage is taken of this astringent quality by incorporating tannin into medicines to check diarrhoea, though it is difficult to get personal views on the reactions of regular wine drinkers in this respect.

Not only is tannin indispensable to the flavour but it is of practical assistance to the production of wine by assisting it to clear easily. This is due to tannin's ability to coagulate proteins in suspension and to precipitate them, causing them to settle out. Occasionally, it is necessary for *collage*, or fining, of wines in order to remove hydrocolloids consisting of pectic substances, gums and proteins that will not settle and are too fine to filter out, their presence causing the wines to appear cloudy. In such cases albuminous substances are added, such as the white of egg, or a gelatine, such as isinglass, which coagulate with the tannin to form a protein–tannin complex, that enmeshes and precipitates these colloidal particles. It is this affinity of tannin for proteins on which the tanning of leather is based, and the employment of tannic acid jelly medicinally to form a coagulated covering over a burn on the skin. During the maturation of wine tannin assists by reacting with basic substances to be deposited as insoluble tannates, so smoothing out the wine in the process.

A number of tannins exist in nature, several being glucosides of gallic acid. Glucose can not only link up with other simple sugars but also, by means of the familiar -O- linkage, with other substances containing -OH groups. In this way plant glucosides are formed which store up for the plant volatile scents and flavours as well as colour pigments. Usually the plant also contains enzymes capable of hydrolising the glucoside, so that crushing and moistening or soaking the leaves or petals liberates the substance compounded with the glucose. Gallic acid, one of the substances that can react with glucose in this way, is chemically an *aromatic acid*, derived from *closed-chain* compounds containing one or more benzene rings; more precisely, it is a *hydroxy-benzoic acid*.

Common tannin, or tannic acid, $C_{14}H_{10}O_9, 2H_2O$, is *di-gallic acid*. It occurs in gall-nuts to the extent of 50%, and can be obtained by treating powdered gall-nuts with a mixture of ether and alcohol. It forms a colourless amorphous mass in its pure state. When boiled in water, it takes up a molecule of water and breaks up into two molecules of gallic acid:

$$\underset{\textit{tannic acid}}{(HO)_2C_6H_2(COOH)—O—(COOH)C_6H_2(HO)_2} + \underset{\textit{water}}{H_2O} \longrightarrow \underset{\textit{gallic acid}}{2(HO)_3C_6H_2COOH}$$

Oak tannin, however, which occurs in oak bark and leaves, is a glucoside of gallic acid, yielding this acid and sugar on hydrolysis, together with ellagic acid and quercitol. Coffee tannin, on the other hand is a glucoside of caffeic acid, and other tannins, although similar in their astringent characteristics, vary in composition in this way.

There is considerable variation of tannin content in fruit juice according to the type and species. That of apples can be as low as 0·05% for Laxton's Superb, and as high as 0·10% for Worcester Permain or even 0·24% for the Kingston Black cider apple. Raspberries are low, with 0·10–0·17%, compared with 0·26% for loganberries. The green gooseberry Keepsake has 0·32%, but the red gooseberry Ironmonger only 0·10%. These figures are given as an indication that there is considerable variation in the tannin content of fruits, but as they refer to unfermented fruit juices, the winemaker must also take into consideration the method he is employing for juice and flavour extraction. The proportion of tannin will be considerably higher where the skins of the ingredients have been left soaking for a day or more, thus giving opportunities for the hydrolysing enzymes to act on the glucosides, and higher still where the initial ferment is conducted on the pulp and skin for a period, because the alcohol formed will assist the extraction.

It is because of this that red grape-wines, fermented on the pulp, contain so much more tannin—and colour pigment—than white or rosé grape-wines. Some ingredients, of course, used for home wine-making, such as flowers, contain no tannin at all, and it may be necessary to add this in the form of commercial grape tannin, or prepared at home from oak leaves or tea. There is a testing procedure available for ascertaining the tannin content of wine. The method used is to add an indigo solution to the wine and then titrate with a solution of potassium permanganate, but as this also involves prior distillation to remove the alcohol, it is really of value only to those with some training in chemistry. The palate remains the indispensable guide to the amount, although it may be useful to compare the 0·1–0.4 g of white grape-wines and the 1·0–3·0 g of red.

Chapter 15 Testing for Acid Content

A. pH Values

Water is not a good conductor of electricity, and pure water which has had its natural salts removed by distillation has a very low electric conductivity ($4{\cdot}3 \times 10^{-8}$ ohm.cm at 18° C). That it provides a path at all for electric current is due to the fact that some of the molecules of water *ionise* or break up into *ions*, particles of moving matter carrying a unitary charge of positive or negative electricity:

	ionisation	
H—O—H ⇋	H^+	and OH^-
water molecule (H_2O)	*positively charged hydrogen ion*	*negatively charged hydroxyl ion*

Each molecule that ionises thus gives a positive hydrogen ion and a negative hydroxyl ion. The proportion of water molecules that ionises is extremely small indeed, hence its weak conductivity; in 1 l. of distilled water there are only $\frac{1}{10000000}$ mole, or 0·0000001 mole, of hydrogen ions, and of course the same amount of hydroxyl ions. (A *mole* is a molecular weight expressed in grams.) As these negative and positive ions balance in quantity, purified water is regarded as being neutral. When substances are added to water these may also ionise, in which case the previous balance of positive and negative ions in the water may be disturbed. If there is an excess of hydrogen ions over hydroxyl ions the solution is *acid*, but if the reverse is true, then the solution is *alkaline*. The relationship of each is such that when one is increased the other is relatively decreased in accordance with the equation:

concentration of H^+ ions $\times$ concentration of OH^- ions = constant value

The characteristic property of an acid is that in water it ionises, or *dissociates*, to give hydrogen ions, e.g.:

HCl ⟶ H^+ and Cl^-

hydrochloric acid

and the characteristic property of an alkali is that in water it ionises to give hydroxyl ions, e.g.:

NaOH ⟶ Na^+ and OH^-

caustic soda

It follows that the degree of ionisation gives us a means of comparing the strength of an acid or base. (A *base* is a compound that reacts with an acid to form a salt and water only; an *alkali* is a soluble base, and therefore all alkalis are also bases.) The strongest will be that which gives the highest concentration of hydrogen or hydroxyl ions in solutions of equal concentration at the same temperature.

We can also measure the acidity or alkalinity of a solution by reference to the hydrogen or hydroxyl ions, though in practice, since one increases in fixed relation as the other decreases, it is usual to speak in terms of hydrogen ions only. The practical difficulty is that alkaline solutions would have so few hydrogen ions that the fraction of moles per litre would be far less than neutral water with its seven places of decimals, and expression in decimals or factional terms would become cumbersome to use.

Sorensen, the Danish scientist of the Carlsberg laboratory, put forward a very neat scheme. As 1000, 100 and 10 may be written 10^3, 10^2 and 10^1, so $\frac{1}{10}$, $\frac{1}{100}$ and $\frac{1}{1000}$ may be written 10^{-1}, 10^{-2} and 10^{-3}, and neutral water with its $\frac{1}{10000000}$ mole of H^+ ions may have this quantity written as 10^{-7} moles. Sorensen decided to refer to the *potential hydrogen-ion* concentration by the letters pH, and to use the index numbers only. Thus $\frac{1}{10000000}$ mole of H^+ ions is written simply: pH 7.

A pH of 6 is therefore ten times as acid as pH 7, a pH of 8 is $\frac{1}{10}$ as acid, and so on. The complete scale runs from 0 to 14, and it will be realised that these figures are not arbitrary ones. For those with a knowledge of logarithms, we can define pH more specifically as *the logarithm to base 10 of the reciprocal of the hydrogen-ion concentration in gram-molecules or moles per litre.*

Table of Hydrogen-ion Concentration in Moles per Litre

			pH	pOH	
1	mole or	10^0	0	14	Acid
$\frac{1}{10}$	mole	10^{-1}	1	13	
$\frac{1}{100}$	mole	10^{-2}	2	12	
$\frac{1}{1000}$	mole	10^{-3}	3	11	
$\frac{1}{10000}$	mole	10^{-4}	4	10	
$\frac{1}{100000}$	mole	10^{-5}	5	9	
$\frac{1}{1000000}$	mole	10^{-6}	6	8	
$\frac{1}{10000000}$	mole	10^{-7}	7	7	Neutra
$\frac{1}{100000000}$	mole	10^{-8}	6	8	Alkaline
		and so to pH 14			

A simple means of measuring the pH of a solution is to employ a set of 'narrow-range indicator papers'. These are in pads about

2 inch × ½ inch, rather like litmus paper, but instead of merely changing to blue or red, they take on a range of colours or shades of colour that can be matched against an indicator card with pH values given for each shade. These 'mixed indicators' are made by employing such dyes as *thymol blue* and *bromocresol green*, which change colour at different pH values. If this creates a picture of chemists in breweries holding soggy pieces of blotting paper up to the light it should be added that these indicator papers are for convenience only; complete accuracy on a commercial scale is obtained by the employment of meters, where a needle moves across a dial calibrated in pH values.

Typical Substances and their pH Value

Substance	pH	
Gastric juice	1·5–2·0	
Lemon juice	2·0–2·5	
Lemon squash	3·0	Acid
Beer	4·0	
Milk	6·6	
Saliva	6·4–6·8	
Distilled water	7·0	Neutral
Blood	7·3	
Drinking tap water	7·4	Alkaline
Pancreatic juice	8·0	
Soapy water	9·0	

Buffer Action

Not all substances ionise to the same degree. The mineral hydrochloric acid ionises almost completely in solution to hydrogen and chlorine ions:

$$HCl \rightarrow H^+ + Cl^-$$

but the organic acetic acid ionises feebly, and it is therefore written:

$$C_2H_4O_2 \leftrightharpoons H^+ + C_2H_3O_2^-$$

The reversible reaction indicates that the hydrogen and acetate ions form a state of equilibrium in solution; if the hydrogen ions are neutralised by the addition of an alkali the unionised portion of acetic acid, the potential source of hydrogen ions, will ionise further to maintain this equilibrium. This explains why it is that 'normal solutions' of hydrochloric and acetic acid each need the same amount of alkali for neutralisation, despite the fact that in terms of pH one is stronger than the other and might appear to need more.

Substances which maintain a pH equilibrium, such as proteins and alkali salts, are said to exert a *buffer action*. Buffer capacity may be defined as a measure of the resistance of a solution to change in its pH when acids or bases are introduced to the solution. If alkali or acid is added to a solution containing buffer substances there will be a resistance to any change of pH values because the equilibrium of ionised and non-ionised molecules will move accordingly in the opposite direction, so that the balance is restored and the concentration of hydrogen ions remains almost the same as before. Dilution of buffer solutions also has very little effect on the pH reading.

Buffer action is of value in such biochemical situations as where it allows the movement of acid through the blood stream of the body without the slight alkalinity of the blood being harmfully varied, the pH remaining fairly constant at 7·3. On the other hand, buffer action means that for winemaking the determination of pH is of little value for the winemaker in arriving at the amount of acid in the must. For example, the pH of lime juice is 1·6–3·2 and for pear juice is 3·0–4·5, so that the figure for samples of each could be 3·0; yet the total *content* of acid for pear juice is about 0·2% and for lime juice about 8·5% an extremely wide variation. The reason is that while in pure aqueous solutions there is a direct correlation between the *active acidity*, measured in pH, and the *amount of acid*, in parts per thousand, this does not necessarily hold true with buffer solutions such as juices and musts of wines. Amerine and Joslyn state in *Table Wines*, 'There is no direct correlation of titrable acidity and pH, undoubtedly because of the widely varying buffer capacity of the must.'

Musts contain organic acids, proteins, salts, pectins and so on, and are altogether complex buffer solutions, so that the winemaker who is interested particularly in acidity from the point of view of its effect on the palate, that is to say, the acid balance of the must in percentage to the whole or in parts per thousand, will not find much guidance from the pH of the must. A recent report by P. M. Duncan shows that an elderberry must which had a pH of 3·7 on a Pye pH meter had in fact an acid content of 16·5 parts per thousand, expressed as sulphuric acid, where an average figure would be 3·5–5·5 parts per thousand. Yet a pH of 3·7 is an acceptable acidity figure for a wine must. In this instance, after dilution with four volumes of water, the pH had moved only to 3·8. S. M. Tritton quotes two wines, each with a pH of 3·6, yet one had 0·58% of acid and the other 0·9%, i.e. about 50% more!

All in all, the home winemaker need not concern himself unduly with the pH of his must. It is true that enzymes are strongly affected by it and have *optimum values*, but those of yeast work inside the cell, and cells possess buffering power in their salts of maintaining a satisfactory pH for their functioning. Yeast itself shows a considerable tolerance to the pH of the medium, working quite well between the very wide range of 3·5–8·0. Most fruit juices are rich in organic acids and satisfactory for yeast activation; a table is given below for comparison of pure fruit juices:

The pH Range of Fruit Juice

Fruit	*pH*	*Fruit*	*pH*	*Fruit*	*pH*
Apple	3·0–4·5	Pear	3·0–4·5	Plum	3·0–4·5
Cherry	3·6–4·4	Peach	3·5–4·0	Apricot	3·0–4·5
Blackcurrant	2·7–3·1	Strawberry	3·0–3·4	Raspberry	2·5–3·1
Grape	2·9–3·9	Orange (sweet)	2·8–4·4	Lemon	2·0–2·5
Lime	1·6–3·2	Grapefruit	2·9–3·6	Pineapple	3·0–3·4

Adapted from A. Pollard 1959, *Organic Acids and Amino Acids in Fruit Juices*

B. Titration

This method of estimating the amount of an acid in solution by neutralising with an alkaline solution of known strength involves the use of an *indicator* added to the acid solution. This is a substance that indicates the completion of a chemical change by a clear change of colour. Some commonly used indicators are *phenolphthalein*, deriving its name from the compounds phenol and phthalic anhydride from which it is formed, *methyl orange* and *methyl red.* These are not employed haphazardly, but according to the general type of reaction taking place, and as phenolphthalein is suitable for the titration of weak acids with strong bases, it is used for testing the organic-acid content of wines. In an acidic or a neutral solution it remains colourless, but a slight excess of alkalinity causes a chemical change to take place in the indicator, accompanied by an immediate switch to a pink colour. It may be objected that since it does not change colour in a neutral solution, too much alkali will have been added when it does change for the result to be accurate. It has to be remembered, however, that a pure compound is not necessarily neutral, but may be slightly alkaline or acidic according to whether it is the salt of strong alkali and weak acid, or vice versa. In the titrating of wine, it is the former, and consequently the compound formed will not be neutral but slightly alkaline and within the pink colour range of the indicator (pH 8·3–10·0). Only a drop or two is needed in the solution being tested, and its value for indicating the *end point* of an acid–alkali neutralisation of this kind is obvious.

The second requirement in titration of wine is an alkaline solution of known concentration or strength, and the standard for this in volumetric analysis is what is termed a *normal solution*. This is a solution one litre of which contains that weight of the substance which combines with or replaces one gram-atom of available hydrogen—its 'gram-equivalent weight'. In fact, this weight is usually the molecular weight in grams, or a simple fraction of it. Thus:

Normal hydrochloric acid (HCl)
Atomic weight of H is 1, of Cl is 35·5
Contains one replaceable hydrogen atom
Therefore N.HCl = 36·5 g of HCl per litre

Normal sulphuric acid (H_2SO_4)—a dibasic acid
Atomic weight of H is 1, of S is 32, of O is 16
Contains two replaceable atoms
Therefore N.$H_2SO_4 = \frac{1}{2}(2+32+64) = 49$ g of H_2SO_4 per litre

Normal acetic acid ($C_2H_4O_2$)
Atomic weight of C is 12, of H is 1, of O is 16
Contains one replaceable hydrogen atom
(only the H in the carboxyl group –COOH is replaceable)
Therefore N.$C_2H_4O_2 = (24+4+32) = 60$ g of $C_2H_4O_2$ per litre

Normal sodium hydroxide (NaOH), an alkali
Atomic weight of Na is 23, of O is 16, of H is 1
Replaces one hydrogen atom of acids
Therefore N.NaOH = (23 + 16 + 1) = 40 g per litre

Concentration need not always be at this strength, and may be varied in terms of normality. Thus a *decinormal* solution, 0·1N, of sodium hydroxide will contain only 4 g l. of distilled water, and a *centinormal* solution, 0·01N, only 0·4 g, whereas a 2N solution will contain 80 g, and so on. As it is desirable in titration that the two solutions used should be of approximately equal normality, it is usual in tests of wine for the alkali used to be of a decinormal solution.

As one litre of a normal solution of an alkali exactly neutralises one litre of a normal solution of an acid, that is displaces all its replaceable hydrogen, it follows that we can work out the unknown normality of any acid solution by measuring carefully how much of a normal solution of alkali is taken to do just this—the phenolphthalein indicator showing when enough has been added to complete the reaction.

This is the process of titration.

The apparatus needed is simple: a pipette, a burette, both calibrated in ml, and a conical flask. Into the burette is put a 0·1N solution of sodium hydroxide (caustic soda), and into the conical flask by means of the pipette 10 ml of the wine to be tested. 10–20 ml of distilled water are added—the exact quantity does not matter, as it is disregarded in the calculation that follows—and two drops of phenolphthalein (a 1% solution in methylated spirits). This solution is then titrated by dropping the sodium hydroxide into the flask, the contents being continuously swirled around, until it just turns pink, and the quantity of sodium hydroxide used is now read off.

Let us assume that this is 8 ml. Then, if we call the wine xN:

$$10 \text{ ml of } x\text{N acid} \equiv 8 \text{ ml of } 0{\cdot}1\text{N alkali}$$

Imagine this as an equation and multiply out each side:

$$10 \text{ ml} \times x = 8 \times 0{\cdot}1$$

$$\text{Therefore } x = \frac{8 \times 0{\cdot}1}{10}$$

$$= 0{\cdot}08$$

The acid content of the wine as a 'normal solution' is thus in this experiment 0·08N. If we wish to express this in terms of a particular acid, then it is multiplied by its gram-equivalent weight. So expressed as sulphuric acid: $0{\cdot}08 \times 49 = 3{\cdot}92$ g l.

For practical purposes, this figure may be regarded as parts per thousand, and therefore the acidity of the wine tested is 3·92 p.p.t. of sulphuric acid. The reason for the choice of sulphuric acid as a standard in wine titration—when it is not, of course, the actual acid of the wine—is that a reasonable indication of acidity may be gained without any calculation, by merely halving the volume of 0·1N–NaOH used. As 8 ml were used in the above test, then $\frac{8}{2} = 4$ p.p.t. of sulphuric acid for the wine tested. This makes possible a very rapid titration calculation for amateurs.

Phenolphthalein as an indicator shows up most clearly in white wines, but even with red wines and musts a change in colour to a bluish-grey will precede the pink hue of the indicator, and act as a warning that the end-point is at hand. It will assist in this respect if more distilled water is added initially, even double the amount. If a fermenting wine is being tested the carbon dioxide will act as carbonic acid, and samples should therefore be boiled briefly to expel the gas in such cases, and then cooled before testing.

As a guide, the following are average acidity figures in p.p.t. sulphuric acid for well-known commercial wines, although there is considerable variation in different samples of the same general types, and it would be advisable for the winemaker to titrate a wine similar to that which he is producing and ascertain for himself its acidity:

Type	*Average acidity (p.p.t.)*	*Type*	*Average acidity (p.p.t.)*
Hock	5·0	Claret	4·0
Moselle	5·0	Sherry	3·0
Burgundy	4·5		

In considering the relative acidity of wines, one must bear in mind that a high alcohol content will mask the acidity of a wine. The same tends to be true of sweet wines, although the fixed acidity in such cases should not therefore be increased to the extent of producing a sweet–sour flavour. Many home winemakers find these figures above rather high for a social wine that is not reproducing a commercial type, and would accept 4 p.p.t. for a sweet wine and 3 p.p.t. for a dry one, but finally it remains a matter for the individual palate.

As a guide to the relative acidity that may be expected of fruits other than grapes, the following table, adapted from *Food and Beverage Analysis* by Bridges and Mattice, for the average p.p.t. acidity of 1 gallon of must prepared from 4 lb of fruit may be useful:

Fruit	*P.p.t. citric acid*	*P.p.t. malic acid*	*P.p.t. other acids*
Apples, crab	0·12	4·08	
Apples, others	—	1·08–3·88	
Apricots, dried	1·40	3·20	Traces of oxalic
Bananas	1·28	1·48	
Blackberries	3·68	0·64	Traces of oxalic
Blackcurrants	9·20	0·20	
Cherries	—	2·24–7·96	
Cranberries	4·40	0·20	4·0 quinic acid
Damsons	—	9·92	
Figs	1·36	Traces	
Gooseberries	Traces	2·0–8·3	
Lemons	15·4	Traces	
Oranges	3·92	Traces	
Peaches	1·48	1·48	
Pears	0·96	0·48	
Pineapples	3·36	0·48	
Raspberries	5·20	0·16	
Rhubarb	1·64	7·08	0·12 oxalic acid
Strawberries	3·64	0·40	
Tomatoes	1·20	0·80	

Acid Adjustment

When titration shows it to be necessary to increase the acid content of a must, and this is more common with home than commercial winemaking, it is usual to make up the deficiency with citric acid, either in the form of lemon juice or acid crystals. This acid has been the standard addition for a very long time, although it has been suggested (Amerine and Joslyn) that some may be converted to acetic acid during fermentation. Of recent years there has been a tendency to use tartaric and malic acids, or even a mixture of all three, on the grounds that these assist flavour and bouquet to an increased extent. The three acids have been arranged (Peynaud) as malic, tartaric and citric in decreasing order of velocity of esterification. Against this has to be set the need for a larger quantity of tartaric acid as an addition, since an amount of this will later be deposited as cream of tartar, and there is also the increased risk of wine haze; in the case of malic acid the possibility exists of a malo-lactic fermentation during storage, again lowering the acid taste of the wine to the palate.

To assist in bringing the titrable acidity to the required level by the addition of the selected acid, the following table may be of value.

Oz per gal	*Parts per thousand sulphuric acid*		
	Citric acid	*Tartaric acid*	*Malic acid*
⅛	0·54	0·51	0·57
¼	1·09	1·02	1·14
⅓	1·46	1·36	1·52
½	2·19	2·04	2·28
⅔	2·91	2·72	3·04
¾	3·29	3·06	3·42
⅞	3·83	3·57	4·09
1	4·37	4·07	4·56
1⅛	4·91	4·58	5·13
1¼	5·46	5·09	5·70
1⅓	5·83	5·43	6·09
1½	6·56	6·11	6·84
1⅔	7·28	6·79	7·60
1¾	7·66	7·13	7·98
1⅞	8·20	7·64	8·55
2	8·73	8·15	9·12

It is not often that acidity requires reducing in home winemaking, but if this should be necessary it can be done by the addition of precipitated chalk, followed by filtering. Each ¼ oz of chalk in 1 gallon of must reduces the acidity by about 1·6 p.p.t. An improvement is the use of anhydrous potassium carbonate, which leaves no after-flavour and forms no sediment, thus avoiding filtering, although it is more expensive to purchase. If 9 oz are mixed with water to make up a volume of 1 pint, ½ fluid oz of this will reduce the acidity of 1 gallon of must by 1 p.p.t. It is interesting to notice in this respect that German vignerons, who are permitted by German wine law to reduce excess acidity in their wines, are now employing the new *doppel salz* or 'Acidex' process in their wines as from 1965.

Chapter 16 Microbiological Sources: Bacteria

Bacteria are the smallest cellular organisms. Although visible under a high-powered microscope, they are so small (the average diameter is about $1\mu = \frac{1}{1000}$ mm) that they appear simply as minute shapes and internal structure cannot be distinguished. The instrument used for their inspection is the electron microscope, which relies on electrons controlled by a magnetic field and not on light focused by glass lenses. A difficulty here is that the object being examined is photographed and not viewed directly by the eye, so that the result presents problems of interpretation in some instances.

Bacterial classification is based, in the first instance, on the shape of the cells: they may be spherical, called *cocci*, long or rod-like, called *bacilli*, curved like a boomerang, called *vibrios* or twisted like a snake, called *spirillia*. The cocci may cluster together, then called *staphylococci*, or form a chain, called *streptococci*:

staphylococci

streptococci

bacillus

vibrio

spirillium

Bacteria have little in common with moulds and yeasts, or indeed with fungi (Eumycetes) in general, but since they multiply by division, they are sometimes classified as *Schizomycetes*, meaning 'splitting fungi'. As they multiply, the cell elongates and then breaks in two across its length, a process named *transverse fission*. Like yeast, the cell consists of protoplasm enclosed in a cell wall, which in turn often has a sufficient layer of slime around it to form a casing called the *capsule*.

Some bacteria form spores that are visible as round swellings in the middle of the cell or at one of its ends:

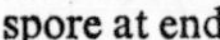

spore at end

spore in middle

and it is these spores that display such remarkable powers of survival in the most adverse conditions over a period of years, waiting for a suitable opportunity to germinate and multiply. Present evidence suggests that bacterial cells contain nuclei, so mitosis is probably involved in cell division; otherwise little is known of the cellular structure.

The terms 'Gram negative' and 'Gram positive' are an outstanding feature of bacterial classification. Gram was a Danish scientist who evolved a method of classifying bacteria by staining them with methyl violet (used to colour methylated spirits) and subsequently treating them with iodine and alcohol. If under this treatment they retain their blue stain they are *gram-positive*, and if not they are *gram-negative*.

From this general introduction we can pass to the bacteria that cause trouble to the winemaker.

A. Lactic Acid Bacteria

Wine and beer yeasts do not contain lactase and are thus unable to utilise lactose. The enzyme is, however, present in certain bacteria which ferment this sugar and produce, not alcohol, but lactic acid as the end-product. When milk turns sour—and this should be carefully distinguished from the effect of putrefactive bacteria on milk, which contaminates it with toxic substances—bacteria, known generally as *Lactobacilli*, have caused the change. Of such species found naturally occurring in this way in milk, *Lactobacillus acidolphilus* and *L. bulgaricus* are frequent, among others. True yoghourt, not that made with synthetic starters supplied by the drug houses, is formed from milk by the employment of *L. bulgaricus* (and *Streptococcus thermophilus*, although this cannot properly be included under *Lactobacilli*), and for the manufacture of lactic acid on a commercial scale in quantity, *L. delbrueckii* is used with beet molasses mash as the medium.

There are numerous species of such bacteria. They are capable of fermenting a variety of substrates in addition to lactose, but a common characteristic is the fermentation of lactic acid as a major, if not the only, product. In the absence of sugars many species can utilise glycerol or a variety of organic acids, including those found in wine.

Two groups of lactic acid bacteria can be distinguished on the basis of their fermentative ability:

1. Homofermentative Types

These are mainly represented by the lactic acid bacteria of milk, and they are not common in fruit juices. Glucose and galactose derived from the hydrolysis of lactose are metabolised to pyruvate by the usual glycolytic sequence, but since the bacteria lack carboxylase activity, the pyruvate is converted to lactic acid instead of acetaldehyde and alcohol. Lactate is virtually the only fermentation product.

lactase
↓
lactose $2C_6H_{12}O_6 \longrightarrow 4CH_3CH(OH)COOH$
monosaccharides (*via pyruvate*) *lactic acid*

2. Heterofermentative Types

While these also produce lactic acid, a number of additional products result from the fact that they also contain carboxylase and aldehyde dehydrogenase. This results in conversion of part of the pyruvate to acetaldehyde, followed by an oxidation–reduction mechanism in which oxidation of acetaldehyde to acetic acid is offset by reduction of pyruvate to lactate, or of a second acetaldehyde molecule to alcohol. In this particular heterofermentative fermentation, therefore, lactate, acetate and alcohol appear as end-products.

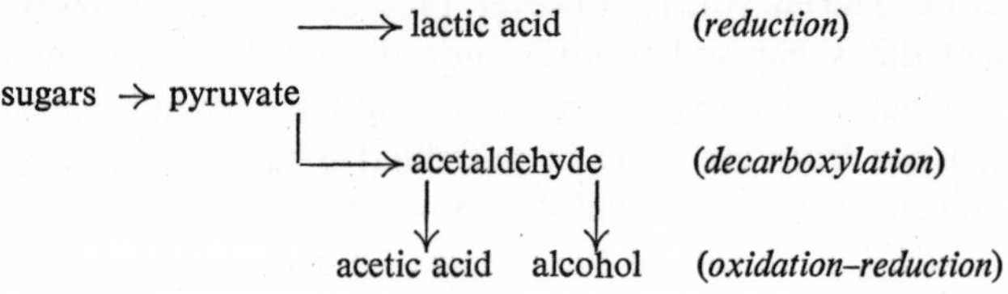

The Effect on Wine

Heterofermentative species of *Lactobacilli* are frequent on fruits, especially when these are overripe, and if these bacteria manage to survive and multiply during the early stages of winemaking disorders may result in the finished wine. Some species, in addition to the products just mentioned, form acetoin and diacetyl, which introduce an undesirable buttery taste. Similarly, a mousy flavour has been attributed to the action of *Bacterium gayoni* and *B. intermedium*. Contamination of tartrate-containing wines, such as those made from grapes, by certain *Lactobacilli* (*L. plantarum* and *Bacillus tartarophthorum*) also results in the appearance of lactic and acetic acids. Pasteur coined the term 'tourne' to describe the bitter and sour flavour thus produced.

In addition to affecting the flavour of the wine, *Lactobacilli* are curious in causing turbidity in wine, which ranges from a cloudiness to thickness and viscosity that in winemaking parlance is usually described as 'ropiness'. The cause may be the failure of dividing cells to separate so that millions of rod-shaped bacilli are linked into chains, appearing as shimmering, silky clouds against a strong light if the wine is agitated, or it may be the formation of a glucose polysaccharide, dextran, together with a gummy slime deposited by the bacteria. Species of *Leuconostoc*, closely related to the *Lactobacilli*, are especially inclined to act in this way. The French term for the disorder is *graisse*.

Prevention is better than cure in dealing with such disorders of taste and appearance, and there are two points here to bear in mind. Firstly, *Lactobacilli* in general are most susceptible to sulphur dioxide, and careful sulphiting of the must therefore reduces danger of infection from this source. Secondly, although they may be described as fairly acid-tolerant, their tolerance of acid varies over a considerable range according to the species. In the commercial production of lactic acid ground chalk is used to neutralise the acid produced in the mash, and in the manufacture of sauerkraut in Germany inoculation by increasingly acid-tolerant species is usual, starting with *Leuconostoc mesenteroides* and progressing through *Lactobacillus plantarum* to *Lactobacillus pentoaceticus*, as each of the earlier species is inhibited by the acid formed. With the majority of species, the ideal pH is 7·5; a pH of 4·5 checks them, and a pH of 3 will destroy them, as Pasteur discovered, if they are in a medium of this acidity for longer than fifteen hours. It follows that adjustment of the acid content of a must offers a means of control, and musts that are deficient in acid should have this added as soon as possible in the winemaking process. The use of an air-lock restricts the risk of a later infection, although, unlike vinegar bacteria, *Lactobacilli*, once established, seem to multiply even more rapidly under anaerobic conditions than they do in air.

Malo-lactic Fermentation

It was Pasteur yet again who noted the change that can take place in the malic acid content of wine. This is an 'acid' acid, and when it is present in any quantity there will be a lack of smoothness on the palate. Wines derived from fruits containing relatively large amounts of malate, such as apples or grapes from colder climates, usually benefit from a malo-lactic fermentation brought about by certain *Lactobacilli*. In the following equation note the loss of one of the acid carboxyl groups (–COOH) and the formation of carbon dioxide, which causes corks to blow unexpectedly in what has been considered a finished wine:

$$\begin{array}{c}CH_2COOH\\|\\CHOHCOOH\end{array} \longrightarrow \begin{array}{c}CH_3\\|\\CHOHCOOH\end{array} + CO_2$$

malic acid *lactic acid* *carbon dioxide*

The classical species for this conversion is *Bacterium gracile*, but there are other *Lactobacilli* spp. such as *L. fermenti* and *L. brevis*, as well as *Micrococcus malolacticus* and *M. acidovorax*. *Lactobacilli* are thermophilic, and warmth encourages this reaction. They also need amino acids and vitamins, so that if autolysed yeast has been left to liberate these growth substances in the wine development of such species will be assisted, and indeed the normal time for malo-lactic fermentation is after the alcoholic fermentation has finished, when the wine has been stored to mature.

B. Acetic Acid Bacteria

It is never very long before the beginner in winemaking encounters the work of these bacteria. About twenty species are recognised, all having an obligatory requirement for aerobic conditions, and all accomplishing the oxidation of alcohol to acetic acid. The reaction is catalysed by the combined action of alcohol dehydrogenase and aldehyde dehydrogenase, acetaldehyde appearing as an intermediate:

$$\begin{array}{l}CH_3CH_2OH\\+NAD\end{array} \rightarrow \begin{array}{r}CH_3CHO\\+NADH_2\end{array} \quad : \begin{array}{l}CH_3CHO\\+NAD\end{array} \xrightarrow{H_2O} \begin{array}{r}CH_3COOH\\+NADH_2\end{array}$$

alcohol *acetaldehyde* *acetaldehyde* *acetic acid*

Free oxygen is involved in securing the ultimate re-oxidation of the reduced coenzymes, but not in the reaction itself.

Fermentation can be defined as an anaerobic partial oxidation in which there is a simultaneous balancing accumulation of a reduced end-product. To describe the action of acetic acid bacteria as 'acetic acid fermentation' is incorrect, unless the term is taken to include also the preliminary fermentation (by yeast) of sugar to alcohol: the expression is nevertheless in frequent use.

As with the *Lactobacilli*, two groups of acetic acid bacteria can be distinguished, for there does not appear to be so much difference between individual species. The first group, termed *Acetobacter*, consists of those types which limit the oxidation to form acetic acid as the end-product, such as *Acetobacter ascendens*, and *A. suboxydens*. Into the second group, termed *Acetomonas*, are placed those types, such as *Acetomonas raucens* and *A. xylinum*, which continue the oxidation process to its ultimate end, converting the acetic acid into carbon dioxide and water. The evolution of gas will not be obvious, since, from the equation, as much O_2 is utilised as CO_2 is produced:

$$\underset{\text{acetic acid}}{CH_3COOH} + \underset{\text{oxygen}}{2O_2} \longrightarrow \underset{\text{carbon dioxide}}{2CO_2} + \underset{\text{water}}{2H_2O}$$

This formulation is, of course, an entirely empirical picture of what takes place. Acetate is presumably converted to acetyl—CoA, which enters the Krebs cycle, resulting in breakdown to CO_2 and formation of reduced cofactors. It is again the re-oxidation of the coenzymes for which atmospheric oxygen is utilised.

The Effect on Wine

If acetic acid bacteria are permitted to multiply freely (this is unlikely, except perhaps in the residue of wine in a neglected bottle, or at the bottom of a cask) they may well form a thick scum containing countless millions of cells. This is the *mycoderma aceti*, a term first used by Persoon as far back as 1822, or in its popular name, 'mother of vinegar'. The species *A. xylinum* is particularly noted for this skin-forming habit; the capsule, or slime-layer, surrounding the cell wall of this bacterium contains a considerable amount of cellulose.

When wine is contaminated to this extent, it rapidly deteriorates to something like soda-water. More usually, however, the oxidation of alcohol is incomplete and acetic acid accumulates. The wine is then *acetified*, with a resulting sharp, piqué taste that one associates with vinegar, and it is useless as a beverage; nothing really effective can be done to remove the damage. Prevention is the best cure here, and the fact that acetic acid bacteria are aerobes provides the clue to this. Care with the primary fermentation in open vessels, where the layer of carbon dioxide over the must is the natural defence against invading bacteria, such as the use of a covering cloth and the avoidance of a delayed fermentation, and the employment of an air-lock for the secondary fermentation in a closed jar or cask, are usually sufficient to prevent the entry of acetobacter. Further it should be noted that *Acetobacter* cannot ferment neat alcohol or even a simple mixture of water and alcohol, for they need nutrients, as do yeasts. Consequently, wine with a low alcoholic content and a layer of disintegrating yeast at the bottom is more prone to attack than wine where the alcohol ratio is higher and frequent racking has left little sediment.

The Fruit Fly

Acetobacter is not infrequently carried by the fruit or vinegar fly, *Drosophila melanogaster*, which appears as if by magic wherever wine is being made.

Although the flies have a very short life-cycle of about six days they breed with extreme rapidity and are claimed to travel distances up to a mile. They are easily distinguished with their red eyes and grey, gauzy wings that project when closed beyond their bodies.

A closely fitting cover over the fermenting vessel in the initial stages of fermentation is usually sufficient to prevent their entry into the must until this is transferred to the cask closed with an air-lock.

C. Butyric Acid Bacteria

The risk of wine being contaminated by bacteria other than those already described is relatively light, but when it occurs the wine is not only spoiled but may even acquire toxic properties. Species of the anaerobic genus *Clostridium* are quick to gain an entry where hygiene is careless and elementary safeguards against infection have not been taken. The species *Cl. pasteurianum* is a nitrogen fixer in the soil, the pathogenic *Cl. botulinum* causes food-poisoning in animals and men, and *Cl. acetobutylicum* is utilised for the manufacture of acetone and butyl alcohol (butanol) from molasses.

In winemaking a special danger is represented by bacteria which produce butyric acid, with its attendant smell of rancid butter. Some of them, notably *Bacillus butyricus*, even accomplish the conversion of lactate to butyrate. Such bacteria ferment starch as well as sugar and are also responsible, for example, for the appearance of butyric acid in dough which has been left too long.

Other bacteria tend to produce foul-smelling amines, often reminiscent of fish (not necessarily *bad* fish), by the decarboxylation of amino acids resulting from protein breakdown. Thus, for example, glycine (aminoacetic acid), the simplest of the amino acids, may undergo decarboxylation in this way to give methylamine, with its fishy odour. Sea-fish, but not fresh-water fish, contain trimethylamine oxide in their tissue, which accounts for their characteristic smell; when left lying about, putrefactive bacteria reduce this to trimethylamine, $(CH_3)_3N$.

Acetamide is another substance which may be produced in the must due to lax hygiene. It may be regarded as an ammonia derivative, where one hydrogen atom has been replaced by an acyl group, and its distinctive and unpleasant smell is reminiscent of mice.

In general, however, bacteria prefer a neutral medium of about pH 7, and putrefactive bacteria do not tolerate the same degree of acidity as *Lactobacillus* and *Acetobacter*. Although it is claimed that species of *Clostridia* can survive, they do not appear to be able to *reproduce* in a must with a pH lower than 5, and the juices of most fruits have a pH between 3 and 4. The early adjustment of musts deficient in acid is an excellent safeguard against contamination apart from the normal process of sulphiting. Pasteurisation, it should be noted, does not destroy spore-forming bacteria.

Chapter 17 Microbiological Sources: Moulds and Yeasts

A. Moulds

The micro-organisms called moulds, once established over the surface of a suitable medium such as jam, are visible to the eye, and perhaps because of this and the fact that we are familiar with them in house and garden, we are less inclined to think of them as sources of wine disorders. Yet they are an ever-present danger to winemaking, and though it is true that the danger largely derives from the use of unsatisfactory raw materials, or from contamination during the early stages when the must is exposed to air and has a negligible alcoholic content, this restriction is balanced by the fact that wine carrying the taint of bitter mould in its flavour is utterly spoilt. No one needs to be reminded of the unpleasant taste of a fruit affected by mould, and we have no way of describing wine so contaminated other than the general description 'mouldy', unless we like to borrow the French term *goût de pourri*. Of the many moulds capable of bringing disaster to wine, we may distinguish two common groups by their colour.

1. Grey and White Moulds

These are the common household 'pin moulds', common enough in the kitchen on a crust of bread that has lain neglected and out of sight for a few days, especially if it is at all damp. There are two main types, the *Mucor* genus, from the Latin word *mucor*='fungus', of which some forty to fifty species may be distinguished, and the *Rhizopus* genus, with some twenty species, from the Greek words *rhiza*='root' and *pous*='foot', the term referring to the characteristic root-like hyphae.

The following table shows their relationship in the plant kingdom:

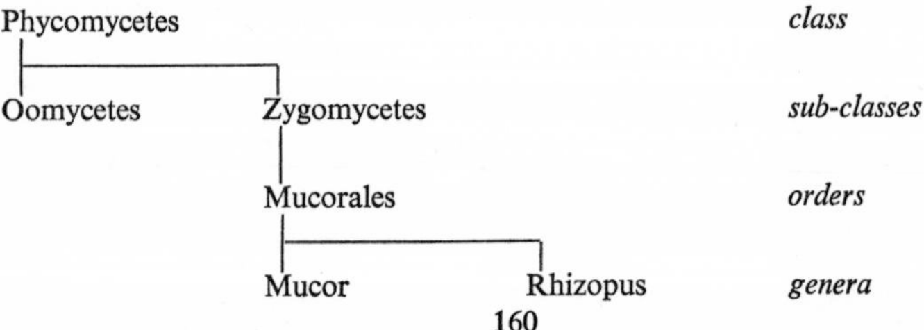

It will be seen that both Mucor and Rhizopus are from the same order of *Mucorales*, the former being its largest genus. Reference should be made to the table of plant life given in the section on yeasts.

Both genera are extremely important to industry, and their value commercially in converting starch to sugar by means of the *diastase* complex of enzymes has been covered in the section on alcohol. They are not at all particular about nourishment: while sugar is very acceptable, they can make do with starch, so that a piece of bread is an excellent substrate. In addition, mineral salts, usually present in wine must, are needed for growth.

The Mucorales multiply by asexual spore formation. On germination the spores give rise to hyphae which penetrate the substrate: the greater part of the resultant mycelium is thus invisible. Spores, however, are formed in globular *sporangia* carried at the tips of erect fertile hyphae, which give the mould its characteristic fluffy appearance. Although there is no recognisable distinction of male and female forms, conjugation may occur between adjacent hyphae to give a zygospore capable of surviving adverse conditions.

2. Blue and Green Moulds

These are a different class of mould altogether. Whereas the Mucorales are included in the Phycomycetes, the blue and green moulds are representatives of Fungi Imperfecti. The two most common genera are *Penicillium* and *Aspergillus*, which are usually responsible for the greenish mould on jam or bread. Particular species are industrially important in the manufacture of blue cheeses, penicillin and citric acid, and in the saccharification of starch as a preliminary to fermentation.

In *Penicillium* the spores are also produced on erect fertile hyphae, but instead of forming sporangia, the hyphae branch at the tips to give a *penicillus*, a Latin word meaning 'an artist's brush', and the spores appear in chains at the ends of the branches. It is the blue and green spores that produce the bright appearance of these moulds. In *Aspergillus* the method of spore formation is similar, but the spore chains are more compact, so that the name taken from the Latin word *aspergillum* = 'a mop for holy water' is a suitable one. Although some species of both genera exhibit conjugation and ascus formation, and would properly be classified as Ascomycetes, it is usual to place the whole group in the Fungi Imperfecti.

The Effect on Wine

As such fungi are mostly aerobes and need atmospheric oxygen, the taint of mould which they bring to wine is conferred in the initial stages of the winemaking process, or even earlier by affected fruit juice of the ingredients. Blue and green moulds are especially common on overripe and damaged fruit in warm seasons. When there is opportunity for them to multiply during the course of winemaking it is usually because for some reason or other the start of the fermentation has been badly delayed, and the spores have time for growth in a still medium which is no longer protected by sulphite because of the long time-lag. In such a case not only do they bring to the must the well-known mouldy taste, but they can also break down the fruit acids, such as malic and citric acids, and leave others behind, such as gluconic acid, thus considerably affecting the flavour of the wine.

The use of ingredients that are sound and clean is the answer to mould infection, and the aid of a cloth over the fermenting vessel during the primary ferment reduces the risk from airborne sources. Moulds are extremely susceptible to the effects of sulphur dioxide, and they give little cause for worry if the must has been sulphited and the ferment that follows is vigorous and undelayed, always provided that the ingredients themselves are uncontaminated.

3. Grape Moulds

Although they do not belong to this chapter, it may be of interest to notice here two types of mould to which the grape vine is particularly prone. One is the powdery mildew, *Uncinula necator*, an Ascomycete, and the other is the downy mildew, *Rhysotheca viticola*, a Phycomycete. Flowers of sulphur control the first, and Bordeaux Mixture the second fungus. That terrible scourge of vines called *Phylloxera vestatrix* is not a fungus at all, but a wingless aphis, related to the common black and greenfly of the garden, which attacks the roots of vines underground. Its control by using native American stocks on which to graft French vines is a well-known story.

From these we can turn to the fungus *Botrytis cinerea*, a grey, ashy mould that is welcomed in the Sauternes district of France and those districts of Germany that produce *trockenbeerenauslese*. Its name is derived from the Greek word *botrys*='bunch of grapes', not because it is found on grapes, for it lives on a wide variety of plants and materials, but because this describes its appearance, as does the Latin word *cinereus*='grey'. When grapes are left unpicked until late in the autumn they start to dry, and the glucose concentration in their juice is increased. Growth of *Botrytis* on and in the grapes results mainly in a decrease in the acid content, with little change in the amount of glucose, so that the quality of the grape juice is improved. No taint of mould is left in the flavour, though it is true that the mycelium of this mould secretes an oxidising enzyme into the juice, called *phenoloxidase* because it acts on the phenolic, or tannic, compounds, and this brings a danger of a late clouding to bottled wine. As the sweet French and German wines depend very much upon the results of Botrytis, it is referred to admiringly as *pourriture noble* or *Edelfäule*—in English = 'noble rot'. It is surely an illustration of the contradictions in nature that another species of this fungus, *Botrytis infestans*, is the potato-blight fungus responsible for the terrible nineteenth-century famine in Ireland.

B. Yeasts

The use of baker's yeast by home winemakers for their alcoholic fermentation may produce a musty, new-bread flavour in the wine, unless care is taken to rack off the wine from the yeast deposit at the close of the fermentation, and brewer's yeast likewise causes a bitter beery off-flavour when used for winemaking. Usually, however, disorders in wine arising from yeasts are due to 'wild yeasts' that have been allowed to establish themselves in the must at an early stage, either being introduced on the skins of the ingredients or being airborne on to the surface of the unprotected liquid. The wild yeasts that are able to produce such disorders in this way have been fully discussed in the section on Yeasts, and it is only necessary here to group them under appropriate titles.

1. Acid Yeasts

The apiculate yeasts produce an excessive amount of acid during their fermentation, and tend also to leave behind them a cider-like flavour in the wine. They are extremely common on fruit, and make a rapid start in fermenting the sugar present in the fruit juice, although the amount of alcohol they can tolerate is very limited. They are usually present in any spontaneous fermentation, that is, one which takes its origin from the yeasts present on the skins of the ingredients. There are two distinct genera: *Hanseniaspora*, the spore-producing yeast, and the imperfect form called *Kloeckera*. The misleading term *Saccharomyces apiculatus* is still sometimes used to describe them.

2. Film Yeasts

The wrinkled skin formed by certain yeasts on the surface of the must is sometimes called 'flowers of wine'. The yeast responsible for this is *Candida mycoderma*, a strongly aerobic yeast capable of utilising the available alcohol, which is duly converted to carbon dioxide and water. Its presence in a must points to gross carelessness in allowing the wine to remain in contact with the air over a considerable period, and normally the skin is found only on wine of low alcohol content that has been left in, say, a half finished bottle of wine, without a cork, The perfect form of this yeast, *Pichia membranaefaciens*, is another oxidative, skin-forming yeast. In the case of sherries, of course, the skin or 'flor' on the must is a desirable feature of fino wines, because it produces the distinctive rancio flavour of this Spanish wine.

3. Slime Yeasts

Associates of moulds are the *Torulopsis* yeasts, regarded as forms of Saccharomyces that have lost their ability to form spores. They are common among wild yeasts and tend to produce sliminess in wines where they are allowed to multiply. Those particularly guilty in this respect because of their capsule-forming properties have now been separated into the genus *Cryptococcus*.

In all cases, wild yeasts, like moulds, can be controlled by sulphiting the must, followed by inoculation with a strongly fermenting wine-yeast starter not later than twenty-four hours after the addition of the sulphur dioxide.

Chapter 18 Chemical Sources

A. Metals

'Casse' is the French term given to wine which contains metallic salts caused by metallic contamination during the making of the wine, or, to a slighter degree, from fungicides sprayed to excess on the vines. The latter may introduce traces of copper, and trace amounts of iron may also be present in grape juice, as this is quite a natural content taken from the soil. The metals iron, copper, zinc, lead, antimony and brass, all dissolve to some extent in an acid liquid, and slight as this may be in winemaking, as few as three parts of metal in a million parts of must are sufficient to cause coloured hazes or cloudiness, together with a roughness and astringency in the taste of the finished wine. The word *casse* is French, taken from *casser*='to break', and when applied to wine affected in this way conveys the idea of 'disintegration'.

The two most common contaminants are iron and copper salts, namely *ferric phosphate* or *ferric tannate*, and *cupric sulphide*. These salts give white, blue and reddish-brown deposits respectively, and even if there are insufficient amounts to cause an early clouding, this eventually occurs, together with darkening of light-coloured wines, as soon as the wine is exposed to oxidative conditions, such as being poured from the bottle into a glass. These troubles are far less likely to affect wine made at home than those produced commercially because of the difficulty for commercial manufacturers in avoiding any contact with metal where wines are made on a scale that involves crushing, pumping and piping with equipment or fittings that incorporate metal in some way. The development of plastic piping and the use of stainless steel and vitreous linings, as well as restriction of spraying towards the time of vintage, have done much in modern times to minimise the danger of disorders from 'metal pick-up'.

In the past, commercial firms could precipitate the metallic complexes so formed by adding *Moeslinger finings*, sometimes referred to as 'blue finings' because of the blue tinge taken on by wine where it is added to excess, and then remove them by filtration. These finings consist of *potassium ferrocyanide*, and because of its reaction with the wine-acids most European countries now regard its addition as hazardous to health and no longer approve of its employment.

If the turbidity is from iron casse the addition of citric acid is helpful in checking further clouding, as iron citrates are then formed which are soluble, but this does not assist in removing any bitterness in the flavour. If this is very slight one may try aeration of the wine, the addition of tannin and finally filtration.

Although the use of earthenware crocks is not common in home winemaking today, one occasionally hears of trouble where the acids of the wine have reacted with the lead of the glaze of an old vessel that has been handed down through the family generations, thus causing lead poisoning in the unhappy imbibers. The practice of using good quality plastic utensils for winemaking in place of the cumbersome crock has everything to commend it.

B. Colloids

In a true solution, or *molecular solution*, the molecules of the substances dissolved are separated, one from the other, by molecules of the solvent, and there is no hindrance to a beam of light shone through the solution. The size of these molecules is usually from one millimicron (1 $m\mu$) downwards. Rather larger than these, say between ten millimicrons and one millimicron, are those molecules which may sometimes appear to give a clear solution, but in fact form a haze which is most clearly visible when light is shone across the liquid, because then the molecules are sufficiently large to scatter the light. The term *colloidal solution* is used in such a situation, and the substances which readily produce the colloidal state are called *colloids*, from the Greek word *colla* = 'glue'.

Usually present in wine after fermentation has ceased are particles of suspended matter from the pulp of the ingredients. The coarsest of these are soon settling at the bottom of the container as the wine begins to clear, but the particles of colloidal size, consisting mainly of gums, proteins and pectic substances, are less inclined to do so. The molecules of the surrounding liquid are in a state of continuous movement, and impacts with these molecules cause the particles to be in a state of constant agitation. This is known in physics as the *Brownian movement*, and it may be observed with a microscope if a beam of light is played across the solution. Consequently, these minute particles can settle only when they are able to come together and agglomerate, so becoming sufficiently large to fall by gravitation to the bottom.

The colloidal particles in wine are surrounded by a layer or film of adsorbed water (*adsorption* is the taking up of one substance at the surface of another), i.e. they are hydrated; in addition, they contain electrical charges, probably formed by the ionisation of free carboxyl groups, with amino and other groups also taking part. It frequently happens where such particles are sufficient to form a haze in the wine that the adsorbed film, together with the electrical charge (the latter is one of the main reasons for the stability of a colloid), interferes with their flocculation, so that the haze obstinately refuses to clear and the wine remains cloudy with the particles permanently in suspension. In such a situation there are for the home winemaker two general methods of treatment.

The first consists of introducing a colloid into the liquid which is of opposite charge, so that there is a mutual attraction between the negatively and positively charged particles, and precipitation then follows as the particles agglomerate. This process is called *collage* or fining, and it consists essentially of adding an agent, such as isinglass, egg-white or agar-agar to the wine, so that these natural poly-electrolytes neutralise the charge on the insoluble matter. In addition, with agents that have imino or amino groups, such as isinglass, hydrogen bonds are established between the $>NH$ and the $-NH_2$ groups of the agent and the $-OH$ groups in the tannin present in the wine, so that a protein–tannin complex is formed which by its coagulation, and consequent precipitation, takes with it the suspended particles. It is important for successful collage, therefore, to ensure that tannin is present in sufficient quantity, and also to conduct an early test with a small amount of the must to determine the amount of fining to add, as an excess will tend to form its own suspension and so increase the haze instead of removing it. An interesting variant from protein and carbohydrate fining agents is bentonite clay. This consists largely of montmorillonite, a hydrated silicate of magnesium formed by the decomposition of volcanic glass under water. A 5% suspension is the usual method of application, and after vigorous agitation the mixture is allowed to settle out. A disadvantage is its marked property of water absorption, causing it to form a large and voluminous deposit, with wasteful results unless filtering follows its use. On the other hand, it leaves no distinctive after-flavour behind, and if grape juice is the main ingredient, or tartaric acid is present in the must, the after deposit of potassium bitartrate is restricted because of bentonite's cation exchange properties.

The second method is to use variation of temperature together with filtration through a medium such as cotton pads or asbestos pulp. Heat causes coagulation of colloidal material; the higher the temperature, therefore, the more efficient the following filtering. About 140° F for two minutes is suggested as practicable. The filter medium acts largely by adsorption of the colloid on the surface of the asbestos fibres, and not by acting as a kind of sieve. An alternative course is to lower the temperature prior to filtering, and in this case 32° F is suggested. The result of this change of temperature is to denature the colloids by dehydration, and although not so efficient as coagulation by heat, the clearing of the wine is effected without loss of body and character. In addition to clarity, removal of suspended colloidal particles means that the trouble of recurrent fermentation is less likely to occur in warm weather, since the nitrogenous matter they contain is no longer available for such yeast cells as remain. Commercial firms incorporate lowering of temperature with the use of centrifugal separators or the conventional horizontal filter-presses with pressure plate filters.

One cannot leave a reference to the heating of wine without an opinion on the resultant character of the wine. Pasteurisation by pumping wine through pipes that have been pre-heated to about 140° F for a period of two to three minutes is quite common with the cheaper wines and Vermouths, and the general result is a brilliant star-clear wine that will throw no deposit nor start to referment. Turin wine intended for Vermouth is kept for ten days at 10° C before heat treatment in a plate pasteuriser. Unfortunately the result is also a dead wine that cannot develop beyond the stage it had reached when pasteurisation took place, and although no doubt the practice is supported by utilitarian reasons, no wine lover can regard it with complete satisfaction. In commenting on the lack of restriction on wine pasteurisation, Morton Shand refers in *A Book of French Wines* to what emerges as 'the mummy of a wine'.

C. Pectins

The juices of fruits and vegetables are contained in the tiny *vacuoles*, or spaces, of the cells. In the walls of these juice-cells, and especially in their middle 'plate' or *lamella*, occurs the substance called *pectin*, combined with cellulose. Pectin has already been analysed chemically in the section on sugars as *methylated polygalacturonic acid*, and reference may be made to this for its chemical composition.

As home jam-makers know very well, the importance of pectin in the jam ingredients lies in its jellifying properties. Provided that suitable concentrations of sugar and acid are present in the mixture, the pectin present will cause the jam to 'gel', or set, firmly. If it is lacking in the ingredients, as it is in some fruits such as strawberries, or if the conditions are unsuitable, the jam remains in a 'runny' state and refuses to set. Commercial forms of pectin are available for addition to jam mixtures where this appears to be necessary, and usually these are obtained from apple pulp, or *pomace*, which is available as a residue in cider factories.

Pectin is essential for successful jam-making, but its presence for similar reasons can cause serious trouble in the production of wine. It may be very unusual for pectin to be present in sufficient amounts to cause much viscosity in wine, although in the presence of calcium salts the pectic acid may be deposited as a jelly-like calcium pectate on occasions. The usual trouble is the formation of a most persistent haze that resists all efforts of fining and filtering, as previously described, to remove it. The explanation is that pectin is a *protective colloid,* and its effect on the colloidal particles is to stabilise them. This means that a stabilised colloidal suspension results and the haze resists effectively attempts to remove it, so that the amateur winemaker is often in despair.

Fortunately, there are pectin-destroying enzymes present in the living fruit-cell, and when the fruits are crushed these enzymes are released and, since regulating mechanisms are no longer in control, they attack the pectin of the ingredients. For this state of affairs to take place, it is essential that the necessary *pectinases* are not destroyed by heat. The use of boiling to extract juices for winemaking is therefore disastrous from the point of view of pectin content, for not only does it kill the pectinases that would normally hydrolyse or break down the pectin but it also extracts more pectin from the ingredients than would be the case with cold water, and so adds to the concentration of this undesirable substance in the wine.

If boiling is unavoidable as with some ingredients used by home winemakers, or if the natural pectinases are considered to be low in quantity, pectinase preparations, that were introduced in the 1920s, are on the market which contain pectin-destroying enzymes that not only degrade the pectin to pectic acid, as do the natural fruit enzymes, but go on to attack the pectic acid and produce soluble galacturonic acid, so that the sticky nature of the first is lost. Some yeasts contain these pectic-acid-destroying enzymes, so that a complete breakdown of pectin is possible by natural means, although the addition of extraneous enzymes leaves nothing to chance. A modern view is that this complete hydrolysis into soluble end-products is not necessary for the removal of pectin's protective role in stabilising colloids, and that clarification occurs after the earlier partial hydrolysis, but this does not affect the efficacy of the pectinase preparations, of course, that contain both *pectinesterase* and *polygalacturonase*

If pectin is suspected as the cause of a stabilised haze a test can be made as with jam-making by adding a small amount of the wine to about four times this amount of methylated spirits, and shaking vigorously together in a bottle. Alcohol precipitates pectin, and consequently, if present, jelly-clots of pectin will appear. Pectinase preparations in the form of liquid or bran can then be employed to degrade the pectin, but it is certainly a better practice to add these to the pulp of the ingredients at the time of starting winemaking. Not only will they then prevent later haze stabilisation but by their early destruction of the pectin-walls, juice will be more freely extracted from the ingredients for fermentation.

D. Oxygen

Controlled oxidation plays an important part in the maturation of wine, but where atmospheric oxygen has free access to the pulp of the ingredients in the early stages of winemaking, excessive oxidation of fruit pigments and phenolics such as leuco-anthocyanins and tannins takes place, with a consequent darkening or browning of the pulp. The result, known as *oxidasic casse*, is that white wines become yellow, and red wines take on an unpleasant rusty tinge, together with a curious aroma and flavour in the finished wine that has been described as 'cooked' and 'caramel-like', so interfering with its vinous quality. This reaction between the oxygen of the air and the chemical constituents of the ingredients is catalysed, or encouraged, by enzymes naturally present in the fruit, called *oxidases*. The timing of the change turns on the acidity of the juice or must. Where the acid content is low, the oxidation is rapid, but it is retarded by an increase in acidity.

It is not often realised that fruits contain intercellular oxygen which, dissolved in solution when they are crushed, is able to react with the pulp and juice as does atmospheric oxygen.

The natural enzymes catalysing this reaction are such as *polyphenol oxidase* and *ascorbic acid oxidase*, which are capable of producing oxidation by molecular oxygen of phenolic compounds and ascorbic acid. The rapid surface darkening of an apple, for example, that everyone has noticed, when it is cut into two is the result of the polyphenol oxidase acting as a catalyst on the chlorogenic acid of the apple tissue.

Commercial firms who market fruit juice in cans and bottles resort to such practices as *deaeration* of the juice or the addition of *glucose oxidase* and *catalase* with a view to exhausting the oxygen that would cause oxidation, but in winemaking the control is sulphiting as soon as possible, followed by a rapid fermentation with no dangerous time-lag in between. These are sufficient safeguards against the deleterious effects of oxidation, especially if the acid content is adjusted where there has been dilution of the natural juices. The results of this form of oxidation on wine should be distinguished from its effect in the presence of *Acetobacter* that break down the alcohol to acetic acid. Oxidised or 'maderised' wine is at least drinkable, but acetified wine has become, or is becoming, vinegar.

Sometimes an excessive amount of atmospheric oxygen is in contact with the finished wine during its maturation, such as when wine, especially white table wine, is stored in small cooperage, say barrels of less than 9 gallons capacity; the ratio of the surface area is high here compared with the capacity of the cask and trouble may follow if the period of maturing in wood is long. Back-oxidation of the alcohol occurs, and an excessive acetaldehyde content results, the flavour being reminiscent of sherry or madeira. As an average, table wines should not contain above 100 p.p.m. of acetaldehyde.

Summary of Disorders

1. *Off-flavours*

Disorder	Cause
Cider-like tartness	Articulate yeasts.
Mustiness	Baker's yeast (possibly).
Bitterness	Moulds. Metallic contaminants. Lactobacilli.
Vinegar-like sharpness	Acetobacter.
Buttery flavour	Lactobacilli.
Cheesiness, fishiness, mousiness	Clostridia and putrefactive bacteria.
Caramel flavour	Oxidasic casse.
Maderised flavour	Excessive aeration.

2. *Appearance*

Disorder	Cause
Surface skin	Acetomonas. Candida yeasts.
Slime	Torulopsis yeasts.
Ropiness	Leuconostoc. Lactobacilli.
Haziness	Colloids, pectins. Metallic contaminants.
Yellowing, rusting	Lactobacilli. Oxidasic casse.

Section 6 The Chemistry of Wine

The name 'organic chemistry' refers back to the early days of chemistry when it was thought that certain substances, especially those composed of hydrogen and carbon, could be made only by living things. As knowledge advanced, so this became progressively untrue, for many came to be made synthetically from matter without life, and therefore the meaning of the term gradually shifted, until today it can be regarded as the study of compounds that contain carbon, for these are essential to all forms of life, rather than the study of substances produced by living organisms.

Carbon atoms have the curious property of linking up very easily with one another or with other atoms to form large strings and circles. The result is that the molecules of organic matter are extremely large. Compare, for example, a molecule of organic sugar containing 22 atoms of hydrogen and 11 atoms of oxygen attached to 12 atoms of carbon, with a molecule of inorganic table salt containing 1 atom of sodium and 1 atom of chlorine.

Another peculiar characteristic of carbon chemistry is the importance of the pattern in which the atoms of the molecule are arranged. It is not enough to know what atoms or how many are used; one must understand how they are put together. Thus diamonds and graphite are both composed entirely of carbon atoms, but the different arrangement of the atoms accounts for their respective hardness and softness.

A study of the formation of the paraffin hydrocarbons is an excellent introduction to organic chemistry, as well as being essential to the understanding of the series of alcohols, and this section therefore commences with a discussion of these.

Chapter 19 The Paraffin Hydrocarbons and Alcohols

A. The Paraffin Hydrocarbons

Hydrocarbon is the name given to a compound composed of carbon and hydrogen atoms only. Enormous numbers of these exist, but we need be acquainted with only four of the *paraffin* hydrocarbons. The paraffin group is distinguished from other hydrocarbon groups in two ways: (1) its members contain the maximum proportion of hydrogen ever found in combination with carbon, and they are therefore often referred to as the *saturated hydrocarbons*; (2) the carbon atoms of its members are attached to one another in the form of an *open chain*, that is the chains are not closed in a ring shape. Such open-chain hydrocarbons are often spoken of as *aliphatic hydrocarbons*, because *fatty* plant and animal substances are included among them.

Methane, CH_4

This is the simplest of all the paraffin hydrocarbons, or, in technical terms, of all the saturated aliphatic hydrocarbons. It has one carbon atom only, and carbon atoms have four 'arms', or bonds, so that its combining capacity, or 'valency', is *four*. The valency of a hydrogen atom is one, so now we can set out the graphic formula of methane showing how the carbon 'grips' the four hydrogen atoms in its molecule:

```
    H
    |
H—C—H   Methane CH4
    |
    H
```

Ethane, CH_5

This paraffin has two carbon atoms, and its graphic formula is:

```
    H  H
    |  |
H—C—C—H   Ethane C2H5
    |  |
    H  H
```

Notice how an 'open chain', if one can use the word 'chain' for two atoms, is forming here by the linkage of carbon atoms. Every other available bond of the carbons is taken up or 'saturated' by the hydrogen atom. The *structural* formula is $CH_3{\cdot}CH_3$.

Propane, C_3H_8

This paraffin has three carbons, and its structural formula is $CH_3{\cdot}\,CH_2{\cdot}\,CH_3$. The graphic formula is as follows:

```
    H  H  H
    |  |  |
H—C—C—C—H   Propane C3H8
    |  |  |
    H  H  H
```

Notice the carbon open chain lengthening, and also that for the first time we have a carbon atom (the middle one) which is attached to *two other carbon atoms.* We shall come back to this point when dealing with alcohols, but at the moment all we need to remember is that such an atom is called a *secondary* atom. Those carbons attached to *one* other carbon atom only (at the end of the chain) are called *primary* atoms.

At this stage we recall the subject of 'isomers' that we encountered in the chapter on sugars. The large number of atoms which make up the hydrocarbon compounds are often capable of being arranged in more than one way, and consequently compounds are possible with the same number of the same kinds of atoms, but with a different structural arrangement of the atoms. Such compounds are called *structural isomers.*

The reader may like to try rearranging the structure of propane above, bearing in mind that the atoms may be swung or rotated in any direction as though the bonds were pivots, so that a true structural isomer is *not* considered to exist if it is formed merely by such rotation. Thus the following is not an isomer of propane:

```
    H  H
    |  |
H—C—C—H
    |  |
    H  | ↙
       |
    H—C—H
       | ↗
       H
```

because by swinging the hydrogen atom SW and the other group NE, we have restored the original formula. In fact, no structural isomer of propane is possible.

Butane, C_4H_{10}

Four carbon atoms are in this molecule, and it has two structural formulae, $CH_3\cdot (CH_2)_2\cdot CH_3$, and $CH(CH_3)_3$, so clearly we shall find structural isomerism arising here for the first time:

```
   H  H  H  H                            H  H  H
   |  |  |  |                            |  |  |
H—C—C—C—C—H normal-butane     H—C—C—C—H isomeric-butane
   |  |  |  |                            |  |  |
   H  H  H  H                            H  |  H
                                            |
                                         H—C—H
                                            |
                                            H
```

The first straight-chain arrangement, *normal-butane*, is called *n*-butane, and the second branched-chain arrangement is called *iso*-butane. This latter is a true isomer, because the graphic formula cannot be changed to the normal straight-chain merely by rotating the atoms, but only by *breaking a bond* and so moving atoms.

As the number of carbon atoms increases, so do the possible isomers, and it has been calculated that the hydrocarbon $C_{13}H_{28}$ has a possible 802 structural formulae, while $C_{20}H_{42}$ could give 366,319 different structures. We leave these to the reader to construct.

One last point to notice, is that the *n*-butane formula has two secondary atoms, and the *iso*butane formula has one *tertiary* carbon atom, attached to *three* other carbon atoms.

Homologous Series

The astute reader will have noticed that each of the paraffins mentioned has one carbon atom and two hydrogen atoms more than the one before. Such a group of compounds is called a homologous series, and paraffins are known up to $C_{62}H_{126}$. To find the number of hydrogen atoms in a paraffin, just double the number of carbons and add two. The members of such a series show a steady gradation in such physical properties as density and boiling point.

The four paraffins above are all gases. Methane is marsh gas, occurring naturally where vegetable matter is decaying under water. It is also found in coal mines, called 'fire-damp', and is the cause of frequent explosions. It burns easily, and makes up about 25–35% of ordinary household gas. If we use natural gas this proportion of methane increases. Ethane, propane and butane are likewise gases. The hydrocarbons above these in the series are liquids, although propane and butane are easily liquefied. Ordinary petrol is mainly a mixture of the six-, seven- and eight-carbon paraffins, called hexane, heptane and octane. The sixteen-carbon paraffins onwards are solids, although some are only just so with very low melting points; paraffin wax is a mixture of paraffins from $C_{18}H_{38}$ to about $C_{43}H_{88}$.

B. The Series of Alcohols

The tremendous difference between paraffin and alcohol is due to the addition of oxygen to the basic composition of hydrogen and carbon. To be more correct, it is the addition of an oxygen atom which attaches itself closely to a hydrogen atom. These two atoms have such an affection for each other, and are held together so closely, that they are called the *hydroxyl group*, and we have encountered them before. An oxygen atom has two bonds, or a valency of two, and a hydrogen atom one, so that there is a free bond for the group to attach itself to other substances:

—O—	+ —H	⟶ —O—H	usually written —*OH*
oxygen valency of two	*hydrogen valency of one*	*linked together, one free bond*	*the hydroxyl group*

When this hydroxyl group attaches itself to one of the paraffin hydrocarbons we have studied, then the substance is an alcohol; it is this group that is the characteristic feature of any formula for alcohol. As the paraffins are saturated hydrocarbons, that is the carbon atoms have a hydrogen atom on each free bond, the hydroxyl group cannot just *add* itself to them, but can link on only by replacing one of the hydrogen atoms and so *substituting* itself.

In this way, another homologous series, this time of alcohols, is formed, parallel to the homologous series of paraffin hydrocarbons. Thus, alcohol made from methane by the substitution of a hydroxyl group for a hydrogen atom is *methyl* alcohol; that made from ethane is *ethyl* alcohol; that from propane is *propyl alcohol.* It can now be seen why chemists could refer to this series as the *hydroxy-paraffins.*

We should note in passing, that there are thousands of different alcohols. Not only are there the very large homologous series of hydroxy-paraffins, but there are also the *polyhydric alcohols* that contain more than one hydroxyl group in each molecule, and *unsaturated alcohols* as well, but these do not concern us here. We are interested only in alcohols that contain in their molecule one hydroxyl group attached to a saturated carbon atom.

Methyl Alcohol, CH_3OH

Because of the distinctive feature of the hydroxyl group, it is usual to write the formula as shown, and not in the form CH_4O. Its graphic formula is:

$$\begin{array}{c} \text{H} \\ | \\ \text{H—C—OH} \\ | \\ \text{H} \end{array} \quad \textit{methyl alcohol, } CH_3OH$$

It will easily be seen in this simplest of all the alcohols with one carbon atom that one of the four hydrogen atoms of methane has been deplaced by the characteristic hydroxyl group. We term this a *primary alcohol*, because this group is linked to a carbon atom that is attached to no other (or not more than one other) carbon.

Ethyl Alcohol, C_2H_5OH

The structural formula is $CH_3{\cdot}CH_2{\cdot}OH$, and the graphic formula is

$$\begin{array}{c} \text{H H} \\ | \;\; | \\ \text{H—C—C—OH} \\ | \;\; | \\ \text{H H} \end{array} \quad \textit{ethyl alcohol, } C_2H_5OH$$

Here again it is easy to see how this group of alcohols is building up on the basis of the series of paraffin hydrocarbons. This is another primary alcohol, as the hydroxyl group is attached to a primary carbon.

Propyl Alcohol, C_3H_7OH

The structural formula of this alcohol with three carbons is $CH_3{\cdot}(CH_2)_2{\cdot}OH$.

$$\begin{array}{c} \text{H H H} \\ | \;\; | \;\; | \\ \text{H—C—C—C—OH} \\ | \;\; | \;\; | \\ \text{H H H} \end{array} \quad \textit{n-propyl alcohol}$$

This is a primary alcohol, because again the hydroxyl is attached to a primary carbon. As there is an isomeric form, however, it is distinguished by the name *n*-propyl alcohol, because it is the normal structure. It is true that the related propane hydrocarbon has no other structural chain, and one may wonder therefore how it is that any other arrangement is possible. The answer is that this particular isomeric alcohol is produced not by the basic variation in the hydrocarbon chain itself but by the point of attachment of the hydroxyl group. Alcohols have more isomeric forms than their related paraffin hydrocarbons, because each of the carbon chains that are possible may have further *substitution isomers.*

```
   H  OH  H
   |   |   |
H—C—C—C—H  iso-propyl alcohol
   |   |   |
   H   H   H
```

This isomeric form is called *iso*-propyl alcohol, and as the hydroxyl group is attached to a secondary carbon, it is a *secondary* alcohol. Its structural formula is $CH_3 \cdot CH(OH) \cdot CH_3$.

No other arrangements are possible, because as said in speaking of hydrocarbons, any such would merely be rearrangements of one of the above two isomers, produced by rotating arms or inverting the formula, and not another true structural pattern.

Butyl Alcohol, C_4H_9OH

It will be remembered that the relative butane hydrocarbon had two isomeric forms, one with an unbranched and one with a branched chain, called respectively normal butane and *iso*butane. Each of these gives rise to two alcohols, because in each case the hydroxyl group can be attached in two positions.

In order to simplify the graphic formulae, only the carbon atoms will now be depicted, and readers should refer back to the hydrocarbons, writing out the full formulae for themselves.

Primary n-*Butyl Alcohol,* $CH_3 \cdot (CH_2)_3 \cdot OH$

```
C—C—C—C—OH
```

This is a primary alcohol because the hydroxyl group is attached to a primary carbon.

Secondary n-*Butyl Alcohol,* $(CH_3)_2 \cdot CH_2 \cdot CHOH$

```
       OH
       |
C—C—C—C
```

This is a secondary alcohol.

Primary iso-*Butyl Alcohol,* $(CH_3)_2 \cdot CH \cdot CH_2 \cdot OH$

```
C—C—C—OH
   |
   C
```

Tertiary iso-*Butyl Alcohol,* $(CH_3)_3 \cdot COH$

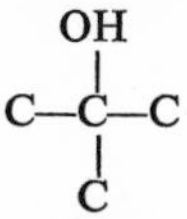

```
  OH
   |
C—C—C
   |
   C
```

This is a *tertiary alcohol* because the hydroxyl group is attached to a carbon atom that is linked to *three* other carbons.

Just as $-CH_2OH$ is typical of a primary alcohol in the formula, showing the hydroxyl attached to an end carbon, so$>$CHOH is typical of a secondary alcohol, and$>$COH of tertiaries.

When writing the names of these four alcohols the names are usually simplified to *n*-butyl alcohol, *sed*-butyl alcohol, *iso*-butyl alcohol and *tert*-butyl alcohol respectively.

Amyl Alcohol, $C_5H_{11}OH$

This alcohol has eight isomers, all of which are known. Four are primaries, three secondaries and one tertiary, resulting again from differences in basic hydrocarbon chain structures and from the position of the hydroxyl group's substitution. Further, three of these isomers contain each an asymmetric carbon atom, and can therefore each occur in two optically active forms and one optically inactive form. (See Optical Isomerism.)

From the point of view of winemaking, two amyl alcohols are of importance because of their occurrence in fusel oil, and there is no reason for proceeding any farther than these in the discussion of the higher alcohols.

Iso-Amyl Alcohol, $(CH_3)_2 \cdot CH \cdot CH_2 \cdot CH_2 \cdot OH$

The first is *iso*-amyl alcohol, often referred to as *fermentation amyl alcohol* because it is the main constituent of fusel oil:

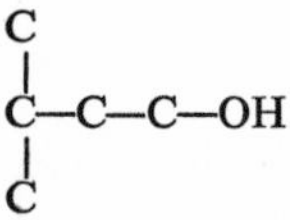

```
C
|
C—C—C—OH
|
C
```

1-Amyl Alcohol, $(C_2H_5) \cdot CH(CH_3) \cdot CH_2 \cdot OH$

As this contains an asymmetric carbon atom, it has optical isomers, two active and one inactive. This particular isomer is laevo-rotatory, and it is therefore often referred to as *optically active amyl alcohol*:

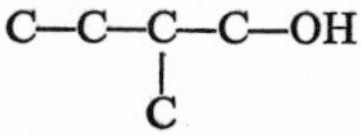

```
C—C—C—C—OH
    |
    C
```

Both of these are primary alcohols.

Chapter 20 The Carbohydrates

The simple sugars and the high polymers of simple sugars, such as starch, glycogen and cellulose, are known collectively as carbohydrates. A distinctive feature of this group is the occurrence of the oxygen and hydrogen atoms, together with carbon, in the proportion of two atoms of hydrogen to one atom of oxygen, just as they occur in water—H_2O. Thus glucose has this pattern—$C_2H_{12}O_6$, and consequently it is a carbohydrate. On the other hand, ethyl alcohol is also made from carbon, hydrogen and oxygen atoms, but its formula, C_2H_6O, does not have this hydrogen/oxygen proportion, so it is not to be classed as one of the carbohydrate group. Incidentally, the term 'carbohydrate' was originally given because at the time it was erroneously assumed from the formula that the molecule consisted of a *hydrated* compound, i.e. that it was a compound which retained a definite amount of water when crystallised from solution. Nevertheless, the term is a useful one, reminding us of the 2:1 proportion of hydrogen and oxygen in the carbohydrate molecule.

A. Monosaccharides

The basic units of the carbohydrate group are the simple sugars known as *monosaccharides*. Their formula always takes the form $C_n(H_2O)_n$, where *n* may be any number from three to ten. To denote the sub-groups of monosaccharides, the suffix *-ose* is used, and hence we have such terms as *pentose* and *hexose*. Most of these sub-groups are of interest to the specialist only.

1. Pentoses

These are monosaccharides with *five* carbons in their molecule, their formula being $C_5H_{10}O_5$, such as *arabinose* from gum-arabic, and *xylose*, or 'wood-sugar' in plants. *Ribose* occurs in plant and animal cells.

2. Hexoses

These are sugars with *six* carbons, and the formula $C_6H_{12}O_6$. Included here are the simple fermentable sugars, well known to winemakers, such as glucose, fructose and galactose, as well as such less-known examples as mannose. They are isomeric sugars, with the same *molecular* formula but varying *structural* formulae, a matter we shall be studying in a moment. It might be interesting to note that in the human liver are *isomerising enzymes* that can convert the sugars mentioned to the common structural form of glucose.

3. Heptoses

Seven carbons distinguish these sugars, and their formula is therefore $C_7H_{14}O_7$. An example is *sedoheptulose*, which plays a role in the photosynthesis of plants.

Optical Isomerism

At this point we should pause awhile to consider in detail what is termed *isomerism* in sugars. One of the most puzzling features of sugars to those who study them for the first time as winemakers is the fact that so many different sugars have an identical formula. Thus, glucose, galactose, mannose and fructose may all be written as $C_6H_{12}O_6$, and in addition there are quite a number of lesser-known hexoses that also use this formula. The reason, as we have already mentioned in passing, is that although these different sugars each contain the same number of the same kind of atoms in each molecule, they differ in the way such atoms are arranged. In other words, while the simplified *molecular* formula is the same for each, we shall see when we write the full *graphic* formula, or *projection* formula, as it is sometimes called, how the pattern varies for each.

Such variation is termed *isomerism*, and the different patterns are called *isomers* or *isomerides*. The type of isomerism that causes so many varying types of sugars is a special form known as 'optical isomerism', or sometimes 'mirror isomerism', because the variations in pattern are those that could be produced by the reflection in a mirror. As an example of this, a person's hands are both composed of four fingers and a thumb; but one hand is the optical isomer of the other, as can be verified by holding the left palm to a mirror and comparing the reflection with the palm of the right hand: they are then identical.

In the same way, there are left-handed and right-handed pairs of substances, referred to as l-*isomers* (l=*laevo*) and d-*isomers* (d=*dextro*). In the previous chapter we met a different form of isomerism that occurred in hydrocarbons and alcohols, called 'structural isomerism', which is not based on this reflected type of variation, but on the fact that certain constitutional atoms form different groups, or the groups are attached to different atoms in the carbon chain. These two forms of isomerism should be carefully distinguished.

Atoms are held together in molecules by electric forces, and different atoms have different powers of attaching themselves to other atoms. One way of understanding these powers is to think of the atoms of different elements as having arms with which they grasp the arms of other atoms to form compounds. A carbon atom has four such arms, or *valency bonds* as they are known. When a different type of atom or group of atoms is attached to each of its four bonds it is said to be an *asymmetric* carbon atom, and substances containing one of these asymmetric atoms will have a pair of *l*- and *d*-isomers; those with two such atoms will have two pairs, and so on. Normally, we show carbon atoms by a two-dimensional drawing, the four valency bonds then being at right-angles to each other, thus:

```
    X
    |
W—C—Y
    |
    Z
```

but this, though convenient, is misleading. It is better to think of the carbon atom as the centre of a tetrahedron, that is a pyramid with three triangular sides and a triangular base—or more correctly, a pyramid with four triangular faces. The four bonds are then the four vertices, or corners, of this:

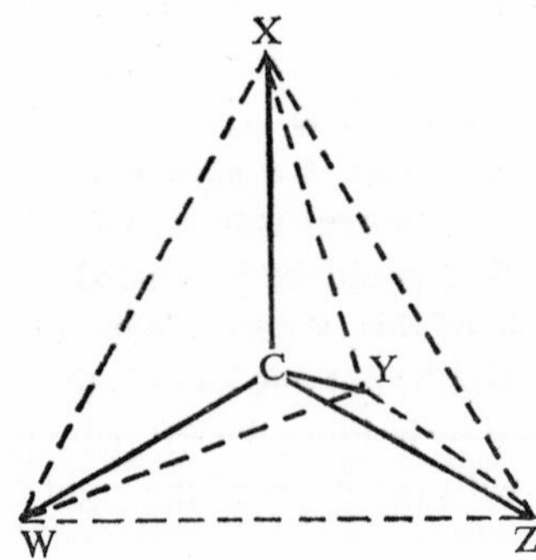

Then these two arrangements on facing page are mirror images of each other, obtained, as it were, by holding the front face to the mirror. It is impossible to *turn* the tetrahedron in any way so that this second form occurs.

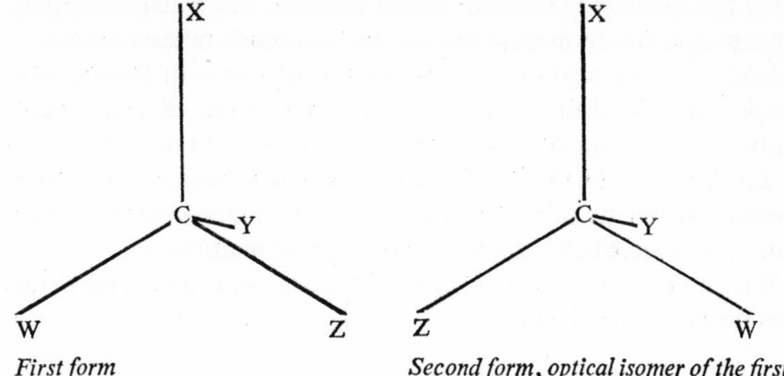

First form *Second form, optical isomer of the first*

On the other hand, if the quadrivalent carbon atom is not asymmetric, but has two or more of the atoms attached of the same type, then it will not have optical isomers. Thus, the following forms are not isomers because by *turning* the imaginary tetrahedron from left to right the second form occurs:

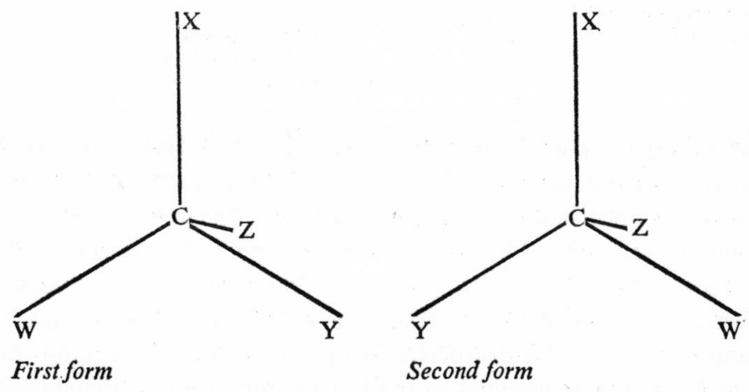

First form *Second form*

If readers find this difficult to visualise from the figures, they should construct a model with a cork as the carbon and matchsticks as the bonds, lettering the ends of the sticks, and the position will then be understood.

With structural isomerism the substances so arising have different chemical as well as physical properties. In the case of optical isomerides, the dextro- and laevo-forms have their chemical and most of their physical properties identical. But they differ physically in one important way: the *direction* of their optical rotation, the *strength* of this being the same for both, and the employment of the polariscope in this respect has already been discussed. Despite such similarity, however, enzymes that use these substances as their substrates never fail to select the correct one of the pair. The zymase complex, for instance, will ferment *d*-glucose, but resolutely refuse to touch its *l*-isomer. As one might expect, therefore, such pairs of isomers vary considerably in their physiological effect, because cells and nerve centres can distinguish between them and respond, or not, accordingly. So, for example, the *l*-nicotine alkaloid from tobacco is extremely poisonous, but its mirror-isomer, *d*-nicotine made synthetically, is much less toxic. Such examples could easily be multiplied.

With an understanding of sugar isomerism, we can pass on to discuss the hexoses in detail.

The Aldoses

There are two main types of hexoses, the *aldoses* and the *ketoses*, and they can be taken separately. The best introduction to the aldoses is the formula of the simplest of the sugars, the *triose*, with three carbons only:

Aldose form of three-carbon sugar: 'aldotriose'	CHO	*The aldehyde group*
	|	
	HCOH	*the hydroxyl group, with asymmetric carbon*
	|	
	CH_2OH	*the alcohol group*

Notice first the top group with the distinctive –CHO of *aldehydes*, that confers the name *aldehyde-sugars* or 'aldoses' to this class of sugars. Chemical-reducing agents can convert this potential aldehyde group or the potential keto-group of a ketose into the corresponding primary or secondary alcohol group. Such sugars are therefore referred to as 'reducing sugars'. The bottom group is an alcohol group, for –CH_2OH is distinctive of a primary alcohol, with the –OH or *hydroxyl group* attached to the last carbon atom of a chain, and consequently the sugars show some of the reactions associated with these alcohols. Finally, the middle group, also containing a hydroxyl, an oxygen and hydrogen atom linked together, has an asymmetric carbon atom; therefore this middle group will show a pair of optical isomers:

Triose sugar formulae	CHO	CHO	*Group showing optical isomerism*
	|	|	
	H—C—OH	HO—C—H	
	|	|	
	CH_2OH	CH_2OH	
	aldose d-*isomer*	*aldose l-isomer*	

By inserting more hydroxyl groups of –CH·OH, that is of the

middle group above, we can produce formulae for the 4, 5, 6, etc. –carbon sugars. We are particularly concerned with hexoses, so now we can set out full projection-formulae of four of the most important hexoses. As such sugars contain four asymmetric carbons each, there are possible $2^4 = 16$ possible isomers, consisting of eight *d*-isomers and eight *l*-isomers, forming eight pairs. All have been produced synthetically, but in nature only three *d*-forms occur, viz. *d*-glucose, *d*-mannose and *d*-galactose, and these are given with d-*gulose*, one of the synthetic types, for comparison:

1.	CHO	CHO	CHO	CHO
2.	HO—C—H	H—C—OH	H—C—OH	H—C—OH
3.	HO—C—H	HO—C—H	H—C—OH	HO—C—H
4.	H—C—OH	H—C—OH	HO—C—H	HO—C—H
5.	H—C—OH	H—C—OH	H—C—OH	H—C—OH
6.	CH_2OH	CH_2OH	CH_2OH	CH_2OH
	d-*mannose*	d-*glucose*	d-*gulose*	d-*galactose*

Notice that the top and bottom groups have not changed, for they do not contain an asymmetric carbon. The fifth group also remains the same for all eight *d*-forms, and its isomer remains the same for all eight *l*-forms; it is therefore the *determinate group* for the *d*- and *l*-classes.

The Ketoses

The two main divisions of monosaccharides are aldoses and *ketoses*, so now we revert to our three-carbon sugar, the aldotriose, in order to examine the arrangement of atoms in its ketonic form. Its formula is:

Ketonic form of three-carbon sugar: 'ketotriose'	CH_2OH	*alcohol group*
	|	
	CO	*ketone group*
	|	
	CH_2OH	*alcohol group*

At once we see that we have here quite a different *structural* arrangement of atoms, so that aldose and ketose sugars of the same molecular formula, in this particular case $C_3H_6O_3$, are *structural* isomers. This particular ketonic three-carbon sugar has no asymmetric carbon, so that it has no optical isomers, as had its aldose counterpart. The middle group, consisting of carbon and oxygen atoms linked together, is known as a *carbonyl* group, and it is distinctive of *ketones*, substances similar to aldehydes, so that sugars containing this group are called *ketonic sugars* or '*ketoses*'.

When we insert more hydroxyl –CH·OH groups as before, we form ketoses with more carbons, and now we shall have asymmetric carbon atoms and therefore isomers. The ketonic hexoses, usually written ketohexoses, six-carbon sugars, have three such carbons, and thus have $2^3=8$ optical isomers, four pairs of *d*- and *l*-forms. The only common one occurring in nature is *d*-fructose (*l*-sorbose also occurs), and its formula is given below:

```
1.        CH2OH
           |
2.         CO
           |
3. HO—C—H      —  containing
           |              asymmetric
4.   H—C—OH  —  carbon
           |              atoms
5.   H—C—OH  —
           |
6.        CH2OH
```

d-*fructose*

It will be noticed that the determinate 5th group has the same optical form in this formula as it had for the *d*-form of the aldoses. It is for this reason that the great sugar chemist, Emil Fischer, decided to term this a d-*form sugar*, although natural fructose rotates the plane of polarisation of light to the left. The *d* and *l* are used with sugars to indicate the spatial arrangement of the determinate group adjacent to the final $-CH_2OH$. The result for the general reader and winemaker is confusing, and these comments should clear up this point, if one remembers that the same sign of rotation does not necessarily mean the same orientation in space.

In conclusion, we can accurately refer to any monosaccharide by naming three features:

1. The number of carbon atoms in the molecule; thus glucose is a *hexose*.
2. The presence of an aldehyde or ketone group in the molecule; thus glucose is an *aldose*, with an aldehyde group.
3. The spatial arrangement of the determinate group; thus glucose is a d-*class* of optical isomer.

B. Disaccharides

As carbohydrates become increasingly complex, they are classified by the number of sets of hexose molecules that their molecules contain. *Monosaccharides* contain one set, although the term is extended to include less common sugars with from three to nine carbons in their molecule; *disaccharides* contain two sets; *polysaccharides* contain three or more sets, although in fact the three principle polysaccharides—starch, cellulose and glycogen—are made up of many more than this minimum of eighteen carbon atoms. It would be more exact to say that these three groups by definition contain three to nine, twelve and eighteen or more carbon atoms respectively, because their formulae are not *exactly* twice or three times that of a hexose: the formula for a disaccharide is $C_{12}H_{22}O_{11}$ and not $C_{12}H_{24}O_{12}$. The reason is that when two hexoses combine, one molecule of water is eliminated:

$$C_6H_{12}O_6 + C_6H_{12}O_6 \text{ less } H_2O \rightarrow C_{12}H_{22}O_{11}$$

So far in order to simplify matters, the graphic formulae of sugars have been set out in line. X-ray crystallography and other evidence, however, have shown that carbon atoms in sugars are arranged spatially not in an open chain but in a ring. It was found, for example, that *d*-glucose existed in two forms with different optical rotation, an α *d*-glucose and a β *d*-glucose. For this to be so, *five*, not four, asymmetrical carbons must exist, and so a six-membered or *pyranose ring* for *d*-glucose has been adopted, and similar cyclic structure for the other hexoses. By way of introduction to the ring formula, a simple form is given below for comparison with earlier examples arranged as open chains:

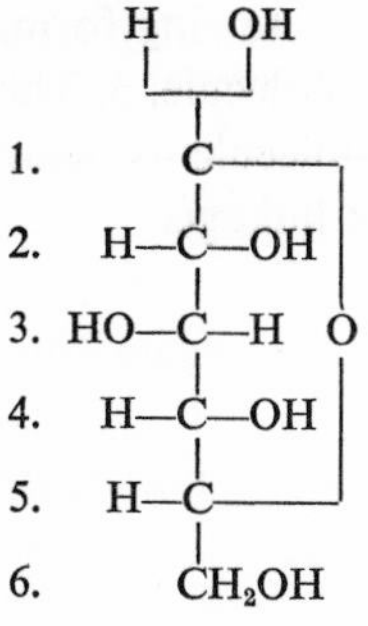

```
          H   OH              HO   H
          |___|                |___|
            |                    |
1.          C——————|             C——————|
            |      |             |      |
2.     H—C—OH      |        H—C—OH      |
            |      |             |      |
3.    HO—C—H       O       HO—C—H       O
            |      |             |      |
4.     H—C—OH      |        H—C—OH      |
            |      |             |      |
5.     H—C—————————|        H—C—————————|
            |                    |
6.        CH2OH                CH2OH
```

α-form of d-*glucose* 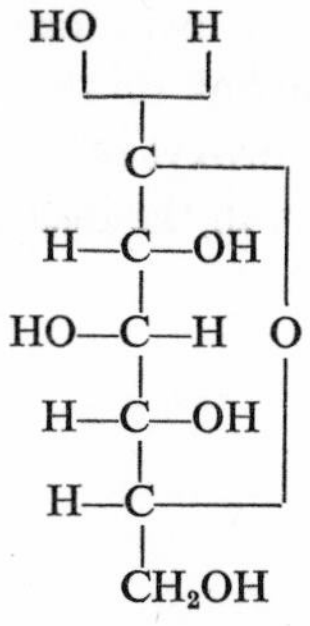 *β-form of* d-*glucose*

Note the 'oxide ring' in which carbon atoms 1 and 5 are linked through oxygen. Carbon atom 1 is now also asymmetrical in this ring formula.

It should be noticed that when fructose is in a combined form it exists as a five-membered or *furanose ring*, and compounds containing it take over some of its instability. Therefore sucrose, which has fructose as one of its components, is much more readily hydrolysed than maltose.

Glycoside Formation

The formation of disaccharides and polysaccharides, or glycoside formation as it is called, is particularly interesting, although it can only be touched on here.

In the ring formula of glucose the carbons are numbered as set out in this 'skeleton formula', with the H's and OH's omitted for the sake of clarity:

```
6C
|
5C—O
|   |
4C  1C glycosidic hydroxyl attached here
|   |
3C–2C
```

Carbon atom number 6 outside the ring represents the end $-CH_2OH$ group of the open-chain formula given earlier. As previously mentioned, such a six-membered form is called a pyranose ring, and a five-membered form is called a furanose ring. The –OH at carbon atom number 1, known as the *glycosidic hydroxyl*, is particularly important because it reacts with the –OH, or hydroxyl groups, of other hexose molecules to form a link. Such hydroxyl groups occur attached to carbons 1, 2, 3, 4 and 6, but in nature such reactions are common only with those at 1, 4 and 6.

To appreciate the structure of a disaccharide, therefore, three points need attention: 1. The nature of the monosaccharide molecules and their arrangement either as a pyranose or furanose ring form. 2. The location of the glycosidic link on the second molecule. 3. The orientation of the glycosidic hydroxyl of the first molecule as being either in the α- or the β-position in the disaccharide linkage.

1. *Maltose*

Here the glycosidic hydroxyl of one molecule of α-glucose reacts with the hydroxyl of carbon 4 of another molecule of α-glucose:

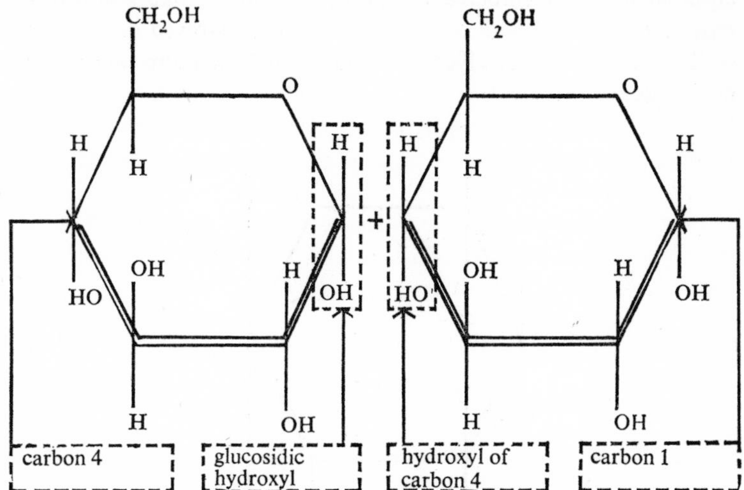

Condensation results, with the formation of H_2O and an oxygen linkage thus:

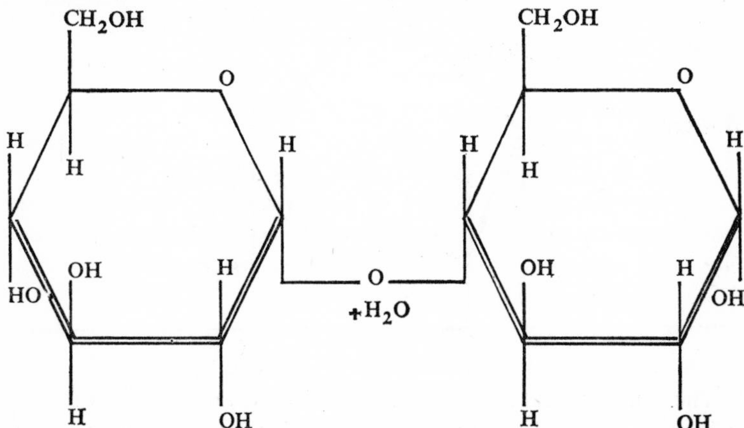

It is the usual practice to omit the carbon atoms for clarity when writing ring formulae, so that they must be imagined at the corner of each figure except at –O–. The full chemical description of maltose is 4-α-glucopyranosyl-α-glucopyranose, and this should now make sense in terms of the linkage.

2. *Sucrose*

In the case of this sugar, the reaction is between the glycosidic hydroxyls of an α-glucose molecule and a β-fructose molecule, the latter in its five-membered or furanose form. Unlike maltose, the reducing groups of *both* components are thus involved in the union, so that there is no free glycosidic radical, and therefore sucrose does not reduce.

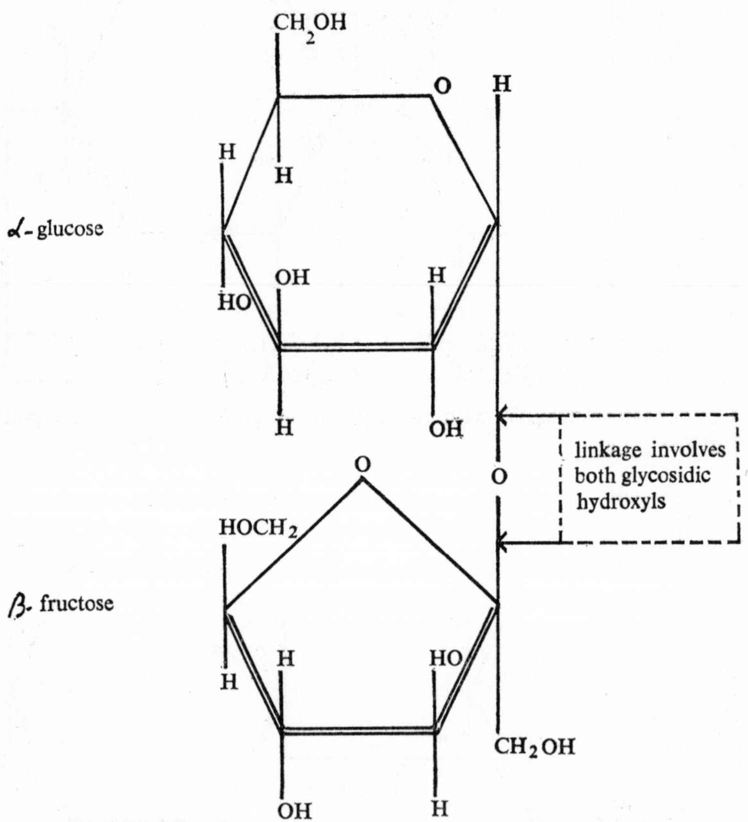

The full chemical name of sucrose is α-glucopyranosyl-β-fructofuranose, exactly descriptive of the units forming this disaccharide.

In similar fashion, hexose units combine to form other disaccharides; thus lactose is 4-β-galactopyranosyl-α-glucopyranose, and the reader should be able to interpret this as signifying that a molecule of β-galactose has linked with a molecule of α-glucose at carbon 4, i.e. end-to-end as with maltose.

C. Polysaccharides

1. Starch

We now know that starch consists of two components, the soluble *amylose* and the less-soluble *amylopectin.* The amylose molecule consists of units of α-glucose with 1–4 linkages like maltose, probably 200–300 of such units end-to-end, and coiled into close spirals with six glucose units in each coil, so forming a long chain. Branching off at carbon-6 are more chains of α-glucose connected through 1–4 linkages, each containing about 30 units and forming the amylopectin molecule. The first constituent gives the well-known deep-blue reaction with iodine, and the latter a reddish-violet colour, so that some idea of the frequency of the amylopectin branches can be gained by the shade of the colour reaction.

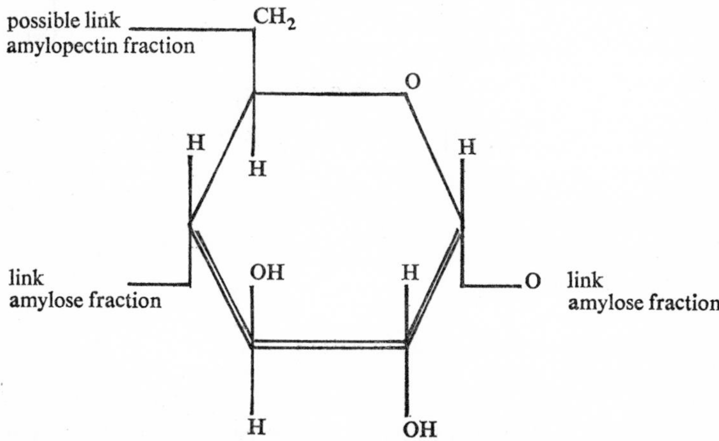

Repetitive unit of starch

2. Cellulose

Each cellulose molecule consists of chains of about 3000–10,000 glucose units. It differs from the starch molecule inasmuch as the units are of β-glucose linked in the particular β-fashion, without branches, and thus amylase that breaks down starch into smaller molecules cannot sever the cellulose linkage. Cellulase, which can bring about hydrolysis, is not common in plants but widespread in micro-organisms.

The following structure of α-*d*-cellobiose, a disaccharide, shows a β-linkage, with the glycosidic hydroxyl of the first molecule in the β-position; this 1–4 linkage should be compared with that of α-*d*-maltose, which has an α-linkage.

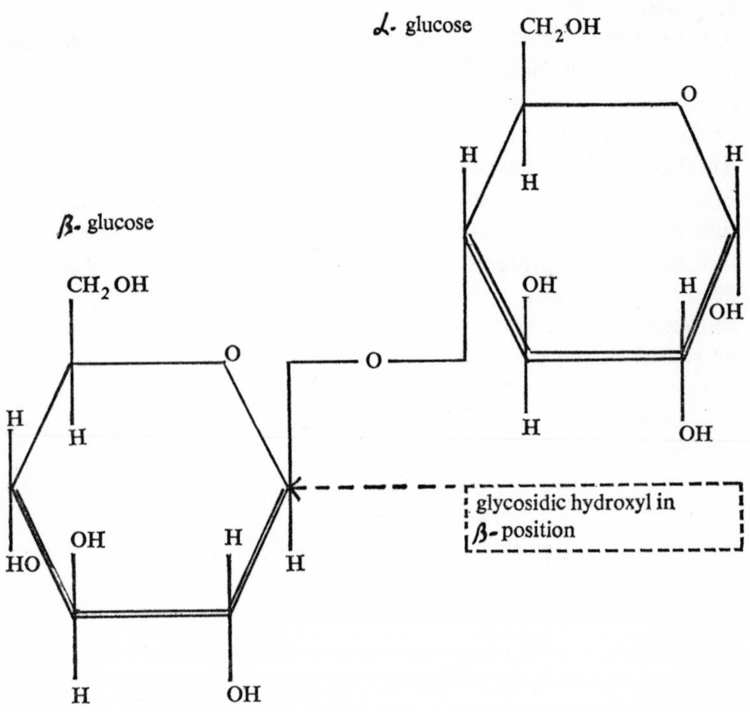

Cellobiose, like maltose, exists in nature.

3. Pectin

The pectin molecule consists of a long chain of units of galacturonic acid, with attached side chains made up of methoxyl groups, residues of methyl alcohol. For biological hydrolysis, two enzymes are necessary; the first of these is pectin pectyl hydrolase, or pectinesterase, which detaches the side chains so that pectic acid and a very small amount of methyl alcohol is formed. The second is poly-galacturonase, which breaks down the pectic acid chain into its units of galacturonic acid.

The polysaccharides present in pectin are galactan and araban; the latter is a polymer of arabinose, a pentose sugar sometimes called pectin sugar, and the former a polymer of galactose.

Occasionally, a wine which has been fermented down to complete dryness will take on a perceptible sweetness after ageing in store. The explanation seems to be a rather high glycerol content together with some residual pentose sugar, which, of course, is non-fermentable by wine yeasts. Glycerol is about one and a half times as sweet as mixed glucose and fructose, and pentose sugar about half as sweet. The softening of the acid and tannin contents during maturation allows this sweetness to come through on to the palate.

The Fermentable Sugars

The Kluyver–Dekker Laws

It has been possible to formulate three rules regarding the regular behaviour of yeasts in their fermentation of carbohydrates. These are known as the *Kluyver–Dekker laws*:

1. A yeast that ferments any other sugar will ferment glucose.
2. If it ferments *d*-glucose, then it will also ferment fructose and mannose.
3. A yeast that ferments maltose will not ferment lactose, and vice versa.

From these it follows that *Saccharomyces cerevisiae* in all its strains, since it is a sugar fermenter, will ferment glucose, fructose and mannose among the simple hexose sugars. One sometimes sees galactose added to these three, but there is a necessary proviso. Not all glucose-fermenting yeasts ferment galactose, and even when they do, like *S. cerevisiae*, they still need a period of acclimatisation unless they have been grown on galactose or have just finished fermenting this sugar. If they have been developed on glucose there is a considerable time-lag when they are transferred to galactose before they adapt themselves to this sugar and start to ferment it.

No other hexoses, apart from these four, are fermented by any yeast, the enzymes being curiously selective in their likes and dislikes of sugars. For example, fructose and glucose are the corresponding keto-hexose and aldo-hexose respectively, and both are fermentable; yet tagatose, which is similarly the keto-hexose corresponding to the aldo-hexose galactose, is unfermentable. The pattern followed is not always what one might expect. Simple sugars, other than hexoses, such as trioses and pentoses, are sometimes fermentable, but usually there is then some resemblance to the hexoses.

The Hydrolase Theory

As regards disaccharides, *S. cerevisiae* ferments sucrose and maltose, but in accordance with law 3 not lactose. In any case, the classical view, known as the *hydrolase theory*, is that all disaccharides are broken down into single hexose units by enzymic hydrolysis prior to fermentation. Such hydrolytic enzymes are specific to one of the hexose units forming the complex sugar, and often such specificity is peculiarly restricted, inasmuch as the enzyme will not necessarily attack these same units when they are combined with a different unit in another type of sugar. In other words, specificity of the enzyme extends, to a certain degree, beyond the hexose unit to the pattern of the molecule as a whole.

Maltose

This restriction is very noticeable with maltase, the enzyme that attacks α-glucosides, of which maltose is an example. It does not, of course, affect β-glucosides, but one might certainly expect it to hydrolyse sucrose, which is also an α-glucoside, although in this case linked to β-fructose, yet it does not do so. On the other hand, it hydrolyses the trisaccharide sugar melezitose, which is composed of α-glucose + β-fructose + α-glucose.

Sucrose

Of the three enzymes known to hydrolyse sucrose, the most common and the one responsible for this in *S. cerevisiae* is invertase. It inverts sucrose into its constituent units of α-glucose and β-fructose by attacking the latter portion of the sucrose molecule. It is particularly specific to a five-member fructose ring, though not necessarily to all of these. For example, it will not hydrolyse this unit in the melezitose sugar above, but it does attack melibiose, the disaccharide mentioned below. Its rate of action more than keeps pace with the fermentation rate of the zymase complex of yeast enzymes, so that the latter is not held up by the need for further sucrose hydrolysis. Incidentally, a pH 4·5 is optimum for both sucrose inversion and fermentation, just as a pH 8 inactivates both, so that they work hand-in-hand, as it were.

Lactose

As lactose, the sugar of mamalian milk, is composed of β-galactose + α-glucose, clearly lactase must specifically attack the β-galactose, since both maltose and lactose contain α-glucose, and from law 3 above a lactose-fermenting yeast does not ever ferment maltose. It does seem strange, however, that both the enzymes lactase and maltase should not occur together in some yeast or other, but apparently no such yeast has yet been found.

Melibiose

This sugar, composed of α-galactose and α-glucose, is interesting because bottom beer-yeasts, as *S. carlsbergensis*, contain the enzyme melibiase, but not the top ones, as *S. cerevisiae*. The enzyme attacks the α-galactose unit.

Raffinose

Composed of α-galactose + α-glucose + β-fructose. The glucose and galactose section are identical with melibiose above, and the fructose and glucose section with sucrose. Invertase separates the fructose, its furanose form being identical with the unit forming sucrose; *S. carlsbergensis* also contains melibiase, so that this yeast goes on to separate the glucose and galactose, and all these three are then fermented. These two sugars, melibiose and raffinose, form a useful test for separating top and bottom yeasts. Neither *S. cerevisiae* nor its variety *ellipsoideus* contains melibiase.

Table of Enzymes

Enzyme	*Attacks*	*Hydrolyses*	*Component Units*
Maltase	α-glucosides	Maltose	α-glucose + α-glucose
Invertase	β-fructosides	Sucrose	β-fructose + α-glucose
Lactase	β-galactosides	Lactose	α-glucose + β-galactose
Melibiase	α-galactosides	Melibiose	α-glucose + α-galactose

Polysaccharides

From the fact that yeast cells do not ferment starch, it is commonly assumed that no enzyme is contained by the cell capable of starch hydrolysis. This is not the case, for expressed yeast juice ferments both starch and glycogen, the animal starch. The explanation for the lack of fermentation is that the starch molecule is too large to enter the membrane of the cell until it has been broken down by external sources of enzymes. The starch-hydrolysing enzymes in yeast are *amyloglucosidases*, attacking both starch and maltose to give glucose, but they cannot achieve this until released by rupture of the cells or by autolysis, as they are not secreted externally. Intact yeast cells do not ferment starch for this reason, although a number of moulds are capable of doing so, and these have been discussed elsewhere under the heading of Starch Saccharification.

The Direct Theory

During the course of this century the well-established hydrolase theory began to look less secure in the face of certain laboratory experiments on the fermentation of sugars. For instance, it was found in certain cases that the rate of fermentation of the component monosaccharides was *slower* than the rate of fermentation of the parent disaccharide; this result did not support the thesis that disaccharide fermentation involved the extra stage of its breaking down by enzymic hydrolysis. And again, the optimum pH for the fermentation of maltose is 4·5, but isolated maltase extracted from the cell and placed in a medium with this pH is quite inactive. This result casts doubts on the view that the fermentation of maltose at such a pH can be preceded by maltase hydrolysis.

These and other experiments were sufficient to encourage an opposition to the hydrolase theory, and those in favour of such a stand became known as 'Willstatter's School', after the name of its leader. Supporters of the conservative attitude, led by the famous Dutch school of biologists and biochemists, rallied strongly to the defence. The answer to the rate of fermentation, they declared, was probably due to the fact that in the laboratory the monosaccharides were supplied externally to the yeast, whereas in practice hydrolysis of disaccharides takes place *inside* the cell, and this latter produced a situation for a faster rate of the component monosaccharide. As regard the pH difficulty, the accumulation of monosaccharides could well inhibit the activity of the hydrolase; in practice, such an accumulation does not occur, because they are fermented away as the breakdown of the disaccharide produces them.

Basically the defence was along the line 'in vivo sed non in vitro'. *In vivo* meaning 'in life' is applied to reactions that take place with living cells; *in vitro* meaning 'in glass', i.e. in the test-tube, refers to experiments that are conducted in the laboratory with isolated extracts and such. It is extremely difficult to demonstrate that an experiment with a hexose and an isolated type of enzyme is identical with the reaction that takes place under normal conditions of life with such complex organisms as yeasts and carbohydrates. Thus a ding-dong battle developed, with the thrust and parry beloved by scientists.

Sufficient time has passed since the 1920s to take a more detached view of the opposing theories, and as is often the case, a sort of compromise seems to be the outcome. The hydrolase theory is too well supported to be disproved, but this does not mean that the direct theory is mistaken, for it is possible for both to exist side by side. The modern view is that another path exists for the fermentation of sugars *in addition* to that which takes place via the hydrolases. We know that glycogen, the polysaccharide formed by yeast cells from sugar very early during fermentation (normally glycogen is the polysaccharide found in animals as distinct from plants), can be broken down by phosphorolysis, and it is therefore possible, though not yet demonstrated conclusively, that phosphorylation in yeasts is the mechanism for the 'direct path'. If this is so, then the disaccharide molecule would be degraded directly to fermentable phosphorylated hexose units without the intervention of hydrolytic enzymes.

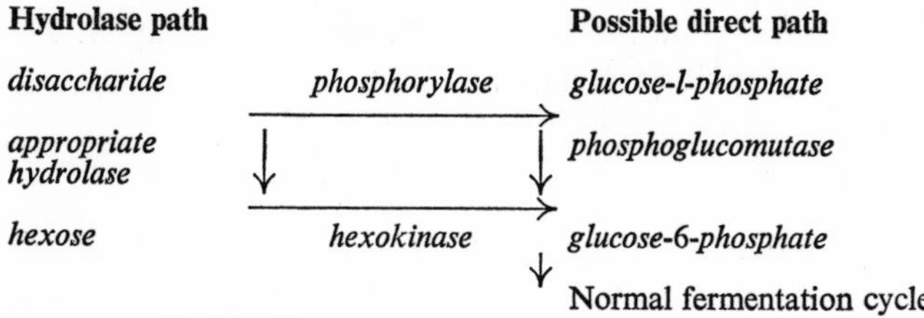

Chapter 21 The Organic Acids

The most distinctive feature of the great majority of organic acids is a group of atoms known as the *carboxyl group*, viz. –COOH, and this will be observed in all the acids mentioned in this chapter. The carboxyl group's graphic formula is:

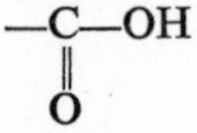

—C—OH
  ‖
  O

and it will be noticed that the *hydroxyl group*, viz. –OH, is also present attached to the carbon atom. It is true that this little group has been said earlier to be distinctive of an alcohol, and it may be queried whether such carboxyl acids are also to be classed as alcohols for this reason. The answer is that such a classification does not apply when the –OH is attached to a carbon atom *carrying an atom other than carbon or hydrogen.* The important point is that in an acid, because of the proximity of this $>C=O$ group, the –OH ionises to $-O^-$ and H^+, whereas in alcohols there is no ionisation.

There is considerable optical isomerism in the acids to be discussed, but as this phenomenon has been covered fully in the previous chapters, it need not be repeated here.

A. The Fatty Acids

Acetic Acid, $C_2H_4O_2$

Acetic acid can be formed by the oxidation of ethyl alcohol, hydrogen being removed and oxygen added:

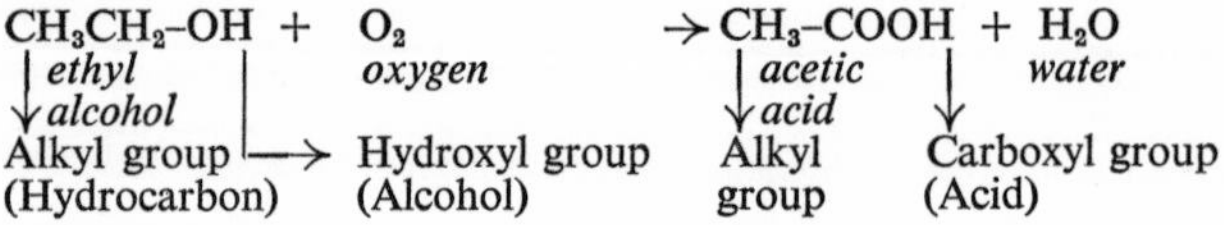

$$CH_3CH_2\text{-}OH + O_2 \rightarrow CH_3\text{-}COOH + H_2O$$

ethyl alcohol ↓ Alkyl group (Hydrocarbon) → Hydroxyl group (Alcohol); *oxygen*; *acetic acid* ↓ Alkyl group; ↓ Carboxyl group (Acid); *water*

It is one of a large series of acids known as *fatty acids* because the acids which many plant and animal fats contain are numbered in this series. They may be regarded as oxidised forms of the homologous series of alcohols, and therefore consist of an alkyl group plus one carboxyl group, as with acetic acid above.

The Homologous Series of Fatty Acids

Carbons	*Related alcohol*	*Acid*	*Structural formula*
C_1	Methyl CH_3OH	Formic	$HCOOH$
C_2	Ethyl CH_3CH_2OH	Acetic	CH_3COOH
C_3	Propyl $CH_3(CH_2)_2OH$	Propionic	CH_3CH_2COOH
C_4	Butyl $CH_3(CH_2)_3OH$	Butyric	$CH_3(CH_2)_2COOH$
etc.	etc.	etc.	etc.

It will be observed that the *alkyl*, or hydrocarbon, group of the acid is 'smaller' each time by CH_2 than that of the related alcohol. One might expect names reminiscent of 'methyl' and 'ethyl' for the first two acids of the series, but the more colourful terms 'formic' and 'acetic' are used because formic acid is found in the ant (Latin *formica* = ant)—and also in nettles, and acetic is the acid of vinegar (Latin *acetum* = vinegar).

Fatty acids are also examples of *monobasic* acids because they have only one carboxyl group in their molecule. This means that there is only one acidic hydrogen atom that can be replaced by a metal atom to form a salt when the acid reacts with a *base*. (Compounds which react with acids to form a salt and water only are called 'bases'.)

As one would expect with acids containing these alkyl groups, there is isomerism in this series of fatty acids, e.g. *n*-butyric acid and *iso*-butyric acid, but the *normal* (unbranched carbon chain) acids are the important ones; the others need not be considered.

B. The Dibasic Acids

Oxalic Acid, $C_2H_2O_4$

Some organic acids have *two* carboxyl groups, and these are then examples of *dibasic acids*. Oxalic acid is the simplest example of this series, its structural formula being:

$$\begin{array}{l} COOH \\ | \\ COOH \end{array}$$

Another example is succinic acid, found as a by-product of the fermentation cycle:

$$\begin{array}{l} CH_2\text{—}COOH \\ | \\ CH_2\text{—}COOH \end{array}$$

This series varies from the fatty acids in forming two types of compounds, depending upon whether one or both carboxyl groups have entered into the reaction. If, for example, in oxalic acid the acid hydrogen in only one carboxyl group has been replaced by a metal we have an *acid oxalate salt,* e.g.

COO | H—not replaced by metal
|
COO | Na—acid H has been replaced

But if that of both groups has been replaced we have a *normal oxalate salt,* e.g.

COONa
|
COONa

And the same thing applies to the esters formed by dibasic acids.

C. The Hydroxy Acids

The two groups of acids dealt with so far are distinguished one from the other by the possession of one or of two carboxyl groups. Now we come to acids which contain *in addition* the well-known hydroxyl group. They are known as the 'hydroxy acids', and are arranged according to the number of such groups possessed by them. They include such important acids as lactic, malic, tartaric and citric acids.

a. Monohydroxy Acids

1. *Lactic Acid,* $C_3H_6O_3$

Lactic acid is a typical hydroxy-monobasic acid, with one hydroxyl and one carboxyl group:

```
                    CH₃
                    |
Asymmetric  ←  C | H | OH →  distinctive single hydroxyl group
carbon atom         |
                    COOH        →  single carboxyl group
```

There is a *structural isomer,* but it is not of importance, and can be disregarded.

The second carbon atom in the formula above is asymmetrical (attached to four different atoms or groups of atoms), and therefore, as we have seen in the previous chapters two *optical isomers,* right- and left-handed forms, exist, viz. *d-lactic* acid and *l-lactic* acid. It is not possible by chemical methods to say which configuration is dextrolactic and which laevolactic acid:

Mirror

CH_3		CH_3
HO—C—H	|	H—C—OH
COOH		COOH

projection formula of the two lactic acids

When lactic acid is obtained by the fermentation of sugar with certain select lactobacillus cultures, such as *Lactobacillus thermophilus*, *l*-lactic acid is formed. The lactic acid that is found in human and animal muscle as a result of the breakdown of glycogen is *d*-lactic acid. When synthetic methods are used in the laboratory to prepare the acid, such as the oxidation of propylene glycol, the result is curious. No effect is made by the resulting acid on polarised light, and this is because synthetic lactic acid is a mixture in equal proportions of the *d*- and *l*-acids, so that these two cancel each other out as regards ray deflection, and the acid is said to be *optically inactive*. It is known consequently as *dl-lactic acid*, a *racemic* mixture. (Berzelius coined this word from the Latin *racemus*='a bunch of grapes', applying it to an inactive form of tartaric acid.)

Chemically, and in most physical properties, the two stereoisomeric acids and the synthetic acid are identical, but all three are easily distinguished by their effect on polarised light.

2. *Malic Acid*, $C_4H_6O_5$

This is a typical *monohydroxy-dibasic* acid, the name indicating that the acid will contain one hydroxyl group, but this time *two* carboxyl acid groups:

COOH —*carboxyl group*
asymmetric carbon atom—C | H | OH —*one hydroxyl group*
CH_2
COOH —*carboxyl group*

Again, just like lactic acid, there is an asymmetric carbon atom in the molecule, so again there are *d*- and *l*-forms, together with a third form consisting of an equal mixture of both, which has no effect on polarised light and is 'optically inactive'.

Mirror

COOH		COOH
HO—C—H	|	H—C—OH
CH_2COOH		CH_2COOH

projection formula of the two malic acids

3. *Citric Acid, $C_6H_8O_7$*

This is a *monohydroxy-tribasic* acid, and therefore it will react to give three series of salts and three series of esters, according to whether one, two or three carboxyl groups have entered into the reaction, as explained before with the dibasic oxalic acid.

```
                         CH₂ | COOH —three
                          |
one hydroxyl group—HO—C—  COOH —carboxyl
                          |
                         CH₂ | COOH —groups
```

There is no asymmetric carbon atom, and so no optical isomers.

b. Dihydroxy Acids

Tartaric Acid, $C_4H_6O_6$

This is a *dihydroxy-dibasic acid,* with *two* hydroxyl groups, as well as two carboxyl groups. The structural formula is:

```
                     1. COOH       — carboxyl group
                         |
two asymmetric — 2.  C | H | OH — two hydroxyl
                         |
carbon atoms   — 3.  C | H | OH — groups
                         |
                     4. COOH       — carboxyl group
```

As there are two asymmetric carbon atoms and each of these can give two stereoisomeric forms, each of which can rotate polarised light to the right or left, the number of optical isomers is therefore $2^2=4$. The *total* effect of the different arrangements of the two H–C–OH groups within the molecule on polarised light will consequently be as follows:

Polarised light deflected by	Group 2 above	Left	Left	Right	Right
	Group 3 above	Left	Right	Right	Left
Total result on polarised light by whole acid molecule		Left rotation	No effect	Right rotation	No effect

Of the four possible optical isomers from this formula, one will give l-*tartaric,* one d-*tartaric,* and two will be similar in being *optically inactive* because their effect on polarised light is cancelled out. In other words, three types of tartaric acid may be expected. In fact, there is a fourth, because like lactic and malic acids, synthetic tartaric acid is a mixture of *d-* and *l*-forms. Thus there are finally four types of tartaric acid:

1. Laevotartaric acid—rotating polarised light anti-clockwise.
2. Dextrotartaric acid—rotating polarised light clockwise.

3. Mesotartaric acid, in which the two groups of atoms capable of isomeric arrangement cancel each other out as regards light rotation in the molecule.
4. Racemic acid (synthetic tartaric acid), in which an equal mixture of laevotartaric and dextrotartaric *molecules* (not atom groups *within* the molecule as in 3) cause the same neutral effect.

The projection formulae of 1, 2 and 3 are:

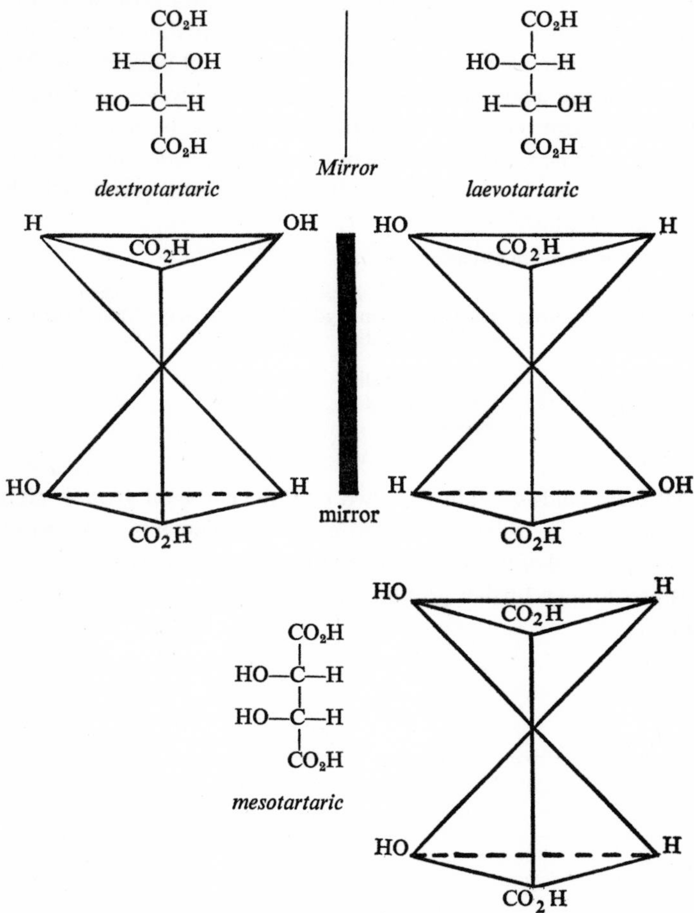

If we imagine polarised light traversing the group of atoms according to the sequence $CO_2H \rightarrow OH \rightarrow C \rightarrow H$ it will be seen in the projection formulae that whereas in one molecule of tartaric acid the direction is clockwise in both upper and lower halves, in the other it will be anti-clockwise in both. In the mesotartaric acid, however, the upper half is anti-clockwise and the lower half is clockwise, so that this acid is optically inactive.

The acid found in nature is the dextro-form, usually as acid potassium tartrate, $KHC_4H_4O_6$. The two inactive forms, mesotartaric acid and racemic acid, result from synthetic methods of preparing tartaric acid. Laevotartaric acid is prepared by separating the two isomerides of racemic acid, and of course dextrotartaric acid can be made in this way too; two optically active acids are thus obtained from an optically inactive acid by this method. It is not possible to obtain *d*- or *l*-tartaric acids from mesotartaric acid, however, because the asymmetry here occurs in two carbon atoms in the *same molecule*, whereas racemic acid is a mixture of *two different molecules*. Similarly, a mixing of *d*- and *l*-tartaric acid solutions will form racemic acid, but never mesotartaric acid.

Classification of Seven Organic Acids

Monobasic			*Dibasic*		*Tribasic*
Fatty acids	Monohydroxy acids	Dibasic acids	Monohydroxy acids	Dihydroxy acids	Monohydroxy acids
Acetic acid	Lactic acid	Oxalic acid Succinic acid	Malic acid	Tartaric acid	Citric acid

All these acids contain one, two or three carboxyl groups, giving the monobasic, dibasic and tribasic arrangements respectively. Further, several of them have in addition one or two hydroxyl groups, thus providing a further grouping into monohydroxy and dihydroxy acids.

Chapter 22 The Biochemistry of Fermentation

The Embden–Meyerhof–Parnas Scheme

As far back as 1815, the Frenchman Gay-Lussac established the equation for anaerobic dissimilation, or fermentation, of glucose as:

$$C_6H_{12}O_6 \rightarrow 2C_2H_5OH + 2CO_2$$

This classic equation sets out correctly the molecular change from glucose to alcohol and carbon dioxide, but it conceals behind its simplicity the complex chain of processes by which this conversion comes about. In this chapter these intermediate reactions will be set out in as simple a manner as possible, although they must by their very nature appear difficult to the non-chemist.

The processes by which cells break down organic molecules in order to obtain energy for living depend upon the battery of enzymes with which they are equipped, and also upon the particular environmental conditions. Types of cells with differing enzymic structure, or the same cells under varying conditions, may follow different schemes in their metabolism. As regards the oxidation of glucose, there are two main pathways, one known as the 'Warburg–Dickens route', and the other as the 'Embden–Meyerhof route'. The former, sometimes called the 'pentose route' because it involves the formation of a five-carbon sugar, is not followed by wine or beer yeasts and need not detain us here. The second of the pathways is the one followed in the fermentation of sugar to alcohol, and therefore the one for close study in this discussion.

Fermentation proper involves the anaerobic metabolism of simple six-carbon sugars (hexoses) to pyruvic acid and ethanol, and preliminary reactions responsible for the breakdown of more complex sugars to hexose will not be considered in this chapter. Hexoses are in any case the major carbohydrate constituents of media such as grape juice. We shall use outline-formulae, based on the skeleton carbon atoms, and omit most of the other atoms in the molecules in order to clarity matters. It must be emphasised that not all of the stages of the Embden–Meyerhof Cycle are given in this outline, but only as many as may seem necessary to convey sufficient understanding of the reactions involved. The numbers used are those of the stages in the full cycle.

These stages can be arranged in three groups:

1. Splitting of glucose into two triose (3-carbon) units.
2. Conversion of triose to pyruvic acid.
3. Conversion of pyruvic acid to alcohol.

Group 1

1. *The Glucose Unit*

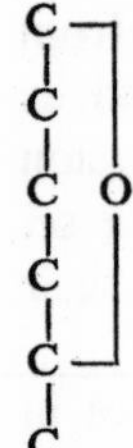

The usual subtrate for fermentation is a hexose sugar (but not *necessarily* glucose).

2. *Glucose–Phosphate*

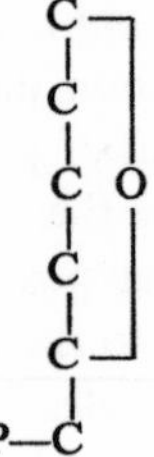

The next stage shows a phosphate attached to the carbon tail, the enzyme *hexokinase* being the catalyst. The vital part that phosphorous plays in glycolysis will be taken separately later.

4. *Fructose-diphosphate*

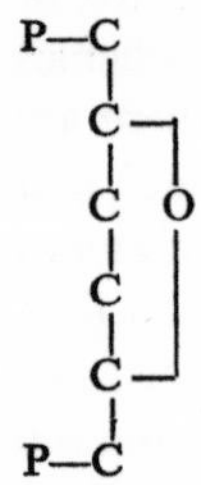

The structural form of the sugar has changed to fructose; only four carbons are now in the ring, and the second tail carbon also has a phosphate attached.

5. *Twin Sugars*

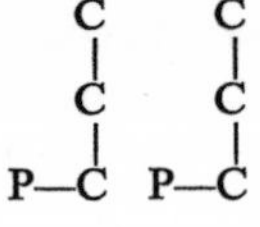

The phosphate ester has split into twin triose phosphates, very simple molecules containing three carbon atoms.

Group 2

8. *A Sugar Acid*

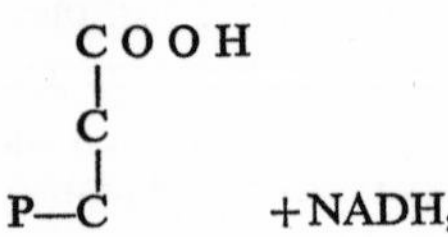

In a reaction involving the transference of hydrogen to NAD, the triose phosphate acquires an acidic carboxyl group (−COOH) and is thereby converted to phosphoglyceric acid.

10. *Pyruvic Acid*

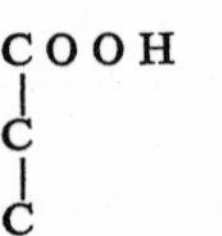

Further reactions, in which the phosphate group is ultimately lost, lead to the appearance of a simple C-3 acid: pyruvic acid.

Group 3

11. *Acetaldehyde*

C + C O O
|
C

The enzyme *carboxylase* catalyses the breakdown of pyruvate to give carbon dioxide (derived from the carboxyl group) and acetaldehyde. This reaction is the main source of the CO_2 evolved during fermentation.

12. *Alcohol*

CH_2OH
|
CH_3 + NAD

The NAD which acted as a hydrogen acceptor in the reaction with triose phosphate now loses the hydrogen to acetaldehyde, which is accordingly reduced to ethyl alcohol.

For those who wish to follow this simplified outline with full formulae, these are now given below. Where a figure is included in the chemical name of the substance, it refers to that number carbon atom in the formula to which the phosphate is attached. Thus *glucose-6-phosphate* indicates that the phosphate is linked to carbon number six.

The Embden–Meyerhof Route

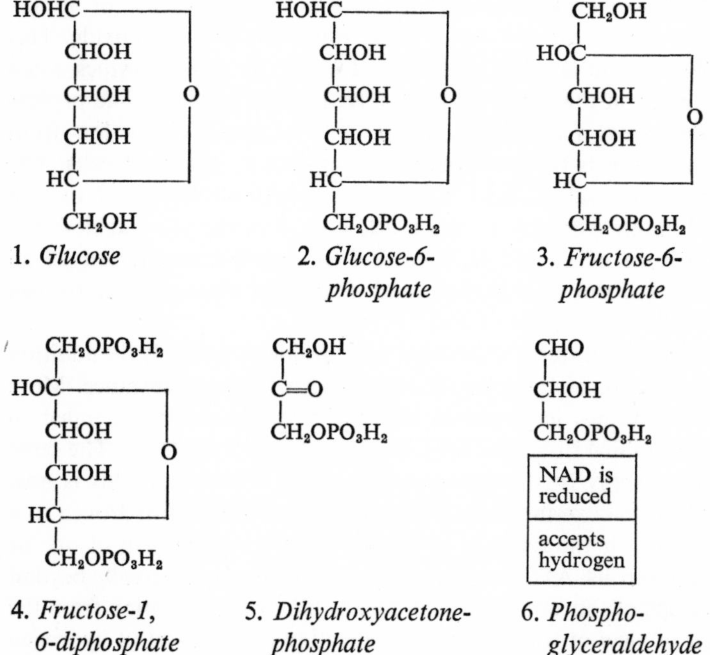

1. *Glucose*

2. *Glucose-6-phosphate*

3. *Fructose-6-phosphate*

4. *Fructose-1, 6-diphosphate*

5. *Dihydroxyacetone-phosphate*

6. *Phospho-glyceraldehyde*

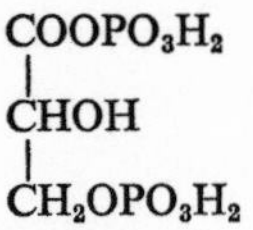

$COOPO_3H_2$ | $CHOH$ | $CH_2OPO_3H_2$	$COOH$ | $CHOH$ | $CH_2OPO_3H_2$	$COOH$ | $C{\cdot}OPO_3H_2$ ‖ CH_2
7. *1, 3-diphospho-glyceric acid*	8. *3-phospho-glyceric acid*	9. *Phospho-enol pyruvic acid*
$COOH$ | CO | CH_3	$CO_2 +$ CHO | CH_3 NAD is oxidised donates hydrogen	CH_2OH | CH_3
10. *Pyruvic Acid*	11. *Acetaldehyde*	12. *Ethyl Alcohol*

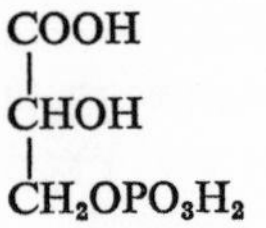

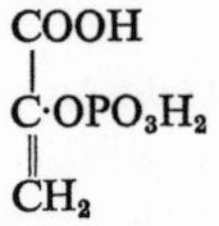

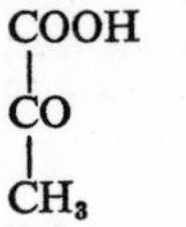

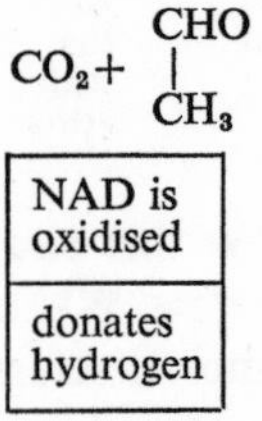

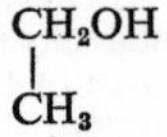

Anaerobic Oxidation

Biological oxidation involving O_2 can be defined as the aerobic oxidation of organic matter by enzymes in the presence of atmospheric oxygen, the net result being water and carbon dioxide. This applies to the aerobic oxidation of sugar by yeast enzymes when oxygen from the air is freely dissolved in the medium. The process is sometimes termed *respiration*, and its basic distinction from fermentation is the consumption of oxygen by the yeast cells. The question naturally arises: in anaerobic conditions when such oxygen is no longer available to the yeast, how does the yeast achieve the breakdown of sugar? Is it correct to regard fermentation as an anaerobic form of oxidation, and if so from whence is the oxygen derived in the absence of air?

Originally the term 'oxidation' was considered to imply the addition of oxygen to, or the removal of hydrogen from, a substance. More modern studies have shown that the term should be extended to include a third situation: the loss of one or more electrons. The same remarks apply to the converse process of 'reduction'. This is now regarded as covering three situations: the addition of hydrogen to a substance, or the removal of oxygen from it, or the gain of one or more electrons. As the subject of electron gain or loss is really beyond the scope of this book, we shall regard biological oxidation from the point of view of winemaking as indicating the removal of hydrogen from, or the *dehydrogenation* of, organic matter.

An important point to bear in mind is that if hydrogen is removed in this way, then it must be transferred to another substance. In other words, oxidation of a substance must be accompanied by reduction of another substance; to balance the *donator* of hydrogen there must be an *acceptor* of hydrogen. In aerobic oxidation this, of course, is atmospheric oxygen, hydrogen removed from the sugar molecule uniting with the oxygen to form water. In so far as sufficient oxygen is available, the sugar will be completely oxidised.

Under anaerobic conditions some hydrogen-acceptor other than oxygen has to be found, and this, as often happens with facultative anaerobes, is another molecule produced during the decomposition of sugar. This means that oxidation of the substrate is incomplete and much of the potential energy of the system remains untapped in the accumulating end-product. Alcohol is a particularly common product of anaerobic plant metabolism.

The hydrogen itself is not left to find its way on its own to an acceptor, but is carried by the appropriate coenzyme. In the case of fermentation, the coenzyme is nicotinamide adenine dinucleotide or NAD. This, as the first known coenzyme, was formerly named Coenzyme I, the 'cozymase' of earlier times. When isolated it was shown to be a compound of the nicotinamide nucleotide and the adenine nucleotide linked by two phosphate groups, and thus a dinucleotide. Individual nucleotides consist of a phosphate radical, a pentose sugar and a nitrogenous base. The base is linked to the glycosidic hydroxyl or reducing group of the sugar, and the phosphate radical to carbon-5 of the sugar ring.

NAD ribose ribose

phosphates

CH_2O P—O—P OCH_2

OH O^-

N^+

$CONH_2$

adenine nicotinamide

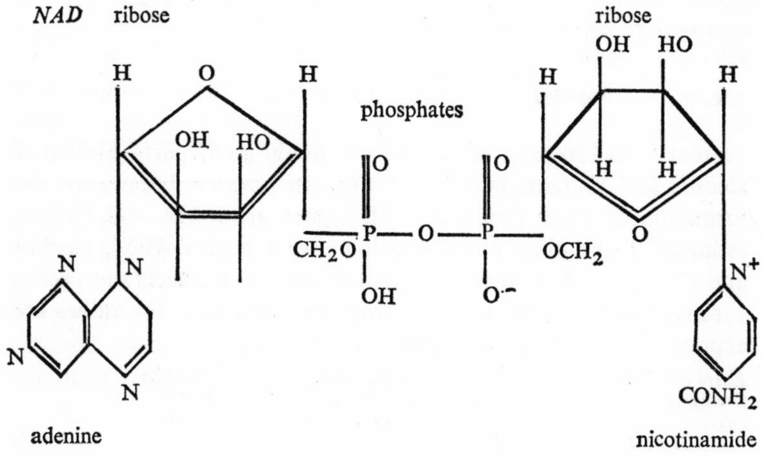

Oxidation of triose phosphate by the action of phosphoglyceraldehyde dehydrogenase results in a transfer of hydrogen to NAD, which is reduced to $NADH_2$. The subsequent reduction of acetaldehyde to alcohol, involving alcohol dehydrogenase, is accompanied by re-oxidation of $NADH_2$ to NAD.

Alcohol fermentation is accordingly a metabolic mechanism which under anaerobic conditions permits the continued growth and multiplication of the yeast until either the substrate is exhausted or the accumulation of alcohol reaches toxic proportions.

Side-products of Fermentation

There exist, however, various alternative mechanisms of anaerobic oxidation. Although of relatively minor importance where yeast metabolism is concerned, the transfer of hydrogen to other acceptors (pyruvic acid, dehydroxyacetone phosphate) is responsible for the appearance of lactic acid and glycerol. Even non-physiological substances such as nitro-compounds and unsaturated alcohols may function as hydrogen acceptors if they are put into the solution.

The extent to which side-products are produced depends on the relative availability of acetaldehyde and the other acceptors, and the activity of the appropriate dehydrogenases: their formation naturally detracts from the yield of alcohol.

If we take Gay-Lussac's formula:

$$C_6H_{12}O_6 \rightarrow 2C_2H_5OH + 2CO_2$$

and substitute the atomic weights, H=1, C=12, 0=16, we get

$$\begin{array}{lll} (72+12+96) & 2(24+5+17) & +\ 2(44) \\ 180 & \rightarrow 92 & +\ 88 \\ \textit{glucose} & \textit{alcohol} & \textit{carbon dioxide} \end{array}$$

So that from 180 parts of glucose we should get 92 parts (51·1 %) of alcohol and 88 parts (48·9 %) of CO_2. In practice, because of the formation of other products, both figures are lower, and Pasteur estimated that complete fermentation gave a yield of 48·6 % alcohol and 47 % CO_2. Modern analyses of fermentation products show some variation in detail, but the following table is typical: the values are expressed as a percentage of the sugar utilised:

Ethyl alcohol	48·4	*Acetic acid*	0·05–0·25
Carbon dioxide	46·5	*Lactic acid*	0·0–0·2
Glycerine	2·5–3·6	*Acetaldehyde*	0·0–0·08
Succinic acid	0·5–0·7		

In general, side-products account for about 4% of the total, although some of them, e.g. the higher alcohols such as amyl alcohol, are not immediately derived from the glycolytic breakdown of sugar. Part of the substrate is also utilised by the yeast in the synthesis of new cellular material.

1. Acetaldehyde

This is the immediate precursor of alcohol. In small amounts it has a fruity smell of apples. The acetaldehyde content of freshly fermented wine is extremely small, but the amount may increase during cask maturation because of a limited back-oxidation of the alcohol. At a concentration of about 100 p.p.m., acetaldehyde starts to affect the flavour of the wine; this is an undesirable development in white table wines, which are then described as *maderised.* Sherry typically contains an acetaldehyde concentration of up to 500 p.p.m.

2. Acetic Acid

This is formed early on in the fermentation by a process called *dismutation*, or a *Cannizzaro reaction*, whereby one molecule of acetaldehyde becomes oxidised to acetic acid, and one molecule is reduced to alcohol. The enzyme *aldehyde mutase* is the catalyst for the reaction:

$$\underbrace{\underset{\textit{Molecule 1}}{CH_3CHO} + \underset{\textit{Molecule 2}}{CH_3CHO}}_{\textit{acetaldehyde}} + \underset{\textit{water}}{H_2O} \rightarrow \underset{\textit{ethyl alcohol}}{C_2H_5OH} + \underset{\textit{acetic acid}}{CH_3COOH}$$

The reaction is, however, reversible, so that as the concentration of alcohol increases it tends to operate in the reverse direction and the ultimate acetic acid content is usually small.

Of course, under aerobic conditions and if the wine is contaminated by *Acetobacter* there will be considerable conversion of alcohol to acetic acid, and the wine will be effectively changed to vinegar.

3. Lactic Acid

Lactic acid, which is formed by direct reduction of pyruvic acid, accumulates throughout the fermentation period, especially if conversion of pyruvate to acetaldehyde and alcohol is restricted, as, for example, when the activity of carboxylase is relatively low.

$$\text{pyruvic acid}\ \begin{array}{c}COOH\\ |\\ CO\\ |\\ CH_3\end{array} + H_2 \rightarrow \text{lactic acid}\ \begin{array}{c}COOH\\ |\\ CHOH\\ |\\ CH_3\end{array}$$

Many *Lactobacillus* spp. contain no carboxylase activity, so that when they appear in a fermentary must the pyruvate they produce in the usual way from sugar is converted not to alcohol but to lactic acid. This may easily constitute a 'disorder' of the wine and the section on 'Disorders of Wine' considers *Lactobacillus* and *Acetobacter* more fully.

4. Glycerine

Glycerol, also known as glycerine, is produced by the transfer of hydrogen to dihydroxyacetone phosphate, an early glycolytic intermediate. The immediate product is glycerol-l-phosphate: this is susceptible to hydrolysis by phosphatase, yielding glycerol and inorganic phosphate. Under normal fermentation conditions the amount of glycerol is small, possibly because glycerol-l-phosphate can also be reoxidised back to triose phosphate and serve after all as a substrate for alcohol formation.

$$\text{(triose phosphate)}\ \begin{array}{c}CH_2OH\\ |\\ CO\\ |\\ CH_2OPO_3H_2\end{array} + NADH_2 \longrightarrow \begin{array}{c}CH_2OH\\ |\\ CHOH\\ |\\ CH_2OPO_3H_2\end{array} + NAD$$

dihydroxyacetone phosphate → *glycerol-1-phosphate*

$$\xrightarrow{H_2O} \begin{array}{c}CH_2OH\\ |\\ CHOH\\ |\\ CH_2OH\end{array} + H_3PO_4$$

glycerol + *phosphate*

Although not an economically sound proposition in normal times, the reaction has been exploited commercially, especially during the First World War, by the use of fermentation conditions whereby acetaldehyde is prevented from functioning as a hydrogen acceptor. In a method developed in this country by Cocking and Lilly, sodium sulphite is added to the fermentation medium. It is a characteristic of aldehydes in general that they react with bisulphite to give 'bisulphite compounds' (hydroxysulphonates); in this way the acetaldehyde is 'trapped' and dihydroxyacetone phosphate takes its place as the next most affective hydrogen acceptor.

	Normal	*Bisulphite*
hexose	$C_6H_{12}O_6$ — — — — — — —	
	↓	↓
triosephosphate	$2C_3H_6O_3$	$2C_3H_6O_3$ *triosephosphate*
	↓	+
pyruvic acid	$2CH_3CO{\cdot}COOH + 2H_2$	$2H_2$ → *glycerol* $2C_3H_8O_3$
	↓	
acetaldehyde	$2CH_3CHO(+2CO_2)$ — — — →	*bisulphite compounds*
	↓	
alcohol	$2C_2H_5OH$	

In America Eoff introduced the use of sodium bicarbonate to give an alkaline fermentation medium. Under these conditions acetaldehyde undergoes a Cannizzaro reaction in which one of two molecules is oxidised to acetic acid and the other is reduced to alcohol. Since uptake of hydrogen is not involved, acetaldehyde fails to function as a hydrogen acceptor and the metabolic pathway is again diverted in the direction of glycerol.

Fermentation in Alkaline Medium

hexose	$C_6H_{12}O_6$ — — — — — — —	
	↓	↓
triosephosphate	$2C_3H_6O_3$	$2C_3H_6O_3$ *triosephosphate*
	↓	
pyruvic acid	$2CH_3CO{\cdot}COOH$	
	↓	+
acetaldehyde	$2CH_3CHO(+2CO_2)$	$2H_2$
	| H_2O	↓
acetic acid	|→ CH_3COOH	$2C_3H_8O_3$ *glycerol*
	↓	
ethyl alcohol	C_2H_5OH	

Energy Transfer

The metabolic conversion of sugar to alcohol, or indeed to carbon dioxide and water, would not seem, at first sight, to be of any special advantage to whichever organism is responsible.

Oxidative breakdown of the substrate is, however, accompanied by the liberation of energy. Whereas in non-biological systems this energy is dissipated as heat, it is a point of fundamental importance that cellular metabolism succeeds in trapping part of the available energy in the form of energy-rich phosphate bonds. Energy held in this way can then be applied to the energy-requiring reactions involved in the synthesis of essential cell constituents.

The yield of energy is obviously greatest under aerobic conditions, when the substrate can be completely oxidised to carbon dioxide and water. Under anaerobic conditions, during alcoholic fermentation for instance, the oxidation is far from complete and the yield of energy-rich bonds proportionately less. The amount of heat generated during biological oxidation is generally slight, but may suffice to raise the temperature of a must a few degrees, and thus incidentally promote the rate of fermentation.

The energy-transfer system which is operative in the glycolytic sequence involves adenosine triphosphate (ATP) as the carrier of an energy-rich phosphate bond. ATP is a nucleotide formed from the base adenine, the pentose sugar ribose and three phosphate groups. Glycolysis is initiated by two successive reactions of the hexose substrate with ATP, to give fructose diphosphate and two ADP (adenosine diphosphate) molecules; but the subsequent reactions result ultimately in the formation of four new ATP molecules by transfer of phosphate to ADP. This means that there is an overall gain of two energy-rich bonds for each glucose unit converted to pyruvate (or alcohol and CO_2).

The three phosphate groups of ATP are given below with the symbol ~ representing an energy-rich bond in accordance with Lipmann's convention:

$$\text{adenosine}—\underset{\underset{\text{O}}{\|}}{\overset{\overset{\text{OH}}{|}}{\text{P}}}\sim\text{O}\sim\underset{\underset{\text{O}}{\|}}{\overset{\overset{\text{OH}}{|}}{\text{P}}}\sim\text{O}\sim\underset{\underset{\text{O}}{\|}}{\overset{\overset{\text{OH}}{|}}{\text{P}}}—\text{OH}$$

Index